Tree Talk

Tree Talk

The people and politics of timber

Ray Raphael

Illustrated by Mark Livingston

Island Press, Covelo, California

Library of Congress Cataloging in Publication Data

Raphael, Ray.
Tree talk: the people and politics of timber.
Bibliography: p. 273
Includes index.
1. Forests and forestry—United States. 2. Forest management—United States. 3. Forest policy—United States. I. Title.
SD143.R32   333.75'0973   81-2835
ISBN 0-933280-10-6   AACR2

Printed in the United States of America

This book was designed and prepared for publication at The Yolla Bolly Press under the supervision of Carolyn and James Robertson. Designed by James Robertson and Diana Fairbanks.

Edited by Michael DiLeo.
Proofread by Juliana Yoder.

The cover illustration was drawn by Mark Livingston from a photograph that is the property of Stihl Incorporated. Used by permission.

*For Nicholas Mattole
and Neil Sylvan*

Acknowledgments

FIRST, I'D LIKE TO THANK all the people interviewed in *Tree Talk*. Without their gracious participation, this book could not exist in its present form. The wide variety of perspectives which they offer adds to the book's vitality. Yet since their viewpoints are varied, nobody is to be held responsible for the words that others might say elsewhere in the book, nor does a personal appearance in *Tree Talk* constitute an endorsement of my own statements and conclusions. Each person speaks only for himself or herself.

In addition, I wish to thank others who offered valuable information about working in the woods: Bob Wazeka, Ron Finne, Andy Johannesen, Hank Behrenst, Cliff Galli, Bryce Gray, Steve Moore, Al Lawrence, Bruce Watson, Mike Bresgal.

During the preparation of the manuscript, I received several helpful suggestions from Gail Lucas, Michael DiLeo, Barbara Dean, and Steve Most. I am gratefully appreciative of their constructive feedback.

Thanks also to Jim and Carolyn Robertson, who took an interest in the project from the beginning and made the necessary connections to facilitate its publication.

Finally, I want to thank the many people who offered practical assistance during the research and preparation of *Tree Talk*: Marie Raphael, Sharon Iveland, Beth de la Fuente, Charlie Sundberg, Sally Weaver, and a host of friends and neighbors.

Contents

Introduction

A FEW hundred years ago, America was covered by virgin forests, hardwoods and softwoods, oaks and maples, firs, pines, and redwoods. Now, the old-growth forests are gone, and we are at the crossroads. Can we replace the old-growth forests and, if so, how? Should we leave nature alone, letting time and biological process heal our wounded woods? Should we take an active approach, engineering man-made forests to replace nature's ancient cathedrals?

We have the opportunity in the next few decades to determine the fate of our future forests. It is an exciting time, one that will require clear thinking, vision, and imagination. *Tree Talk* is my attempt to give a rational analysis of the directions we are currently taking — and to suggest some new ideas of my own.

I am not a professional forester. I am not a logger. But I am a small timber owner, and I do live in the woods. Perhaps I am overstepping my bounds by evaluating the work of professionals; but in the field of forestry, nobody seems to have all the answers, not even the pros. Truly, it is hard to see the forest for the trees.

A forest is different things to different people. A forest is the habitat of wild animals, the dwelling place of magic and enchantment, the residence of building materials and fuel. Personal perspectives loom large in the study of forests. Each individual perspective is significant, yet each is limited, too. There is as much variety among people's perceptions of the forest as there is within the forest itself. To reflect this diversity, I have structured *Tree Talk* around personal interviews with men and women who have firsthand knowledge of the woods. In these pages you will meet cat loggers and horse loggers, naturalists and timber company executives, old-time woodsmen and young pioneers. The book takes the form of a documentary film: the running narrative is punctuated by individual portraits, which are intended to personalize the issues, to translate both the academic and political aspects of forestry into human terms.

In part, *Tree Talk* serves as a popular forum in which the contemporary issues of forestry can be debated. My object is not to rehash arguments for and against wilderness classification, an issue that has attracted considerable attention and led to bitter polarization. Instead, I concentrate on the basic, day-to-day problems of commercial forestry. The wilderness controversy might make the headlines, but the major battlegrounds are the cutover but potentially productive lands that have already been allocated for commercial utilization. How can we provide the forest products we need on a sustained-yield basis? How can we extract timber from the woods, while simultaneously preserving a healthy environment? How can we employ people in safe, stable, but personally challenging occupations within the forest products industry?

It would be fine if common sense alone could answer these questions, but what seems like common sense to one person may be heresy to someone else. Belief systems of religious intensity have come to dominate the study of the woods. A very limited number of experiments, for instance, have been conducted on the controversial phenoxy herbicides, yet this finite body of data has convinced some people that the chemicals are absolutely safe, while simultaneously convincing others that they are extremely dangerous. Predispositions — or should we call them prejudices? — determine the way we evaluate factual information. Although foresters may aspire to scientific objectivity, their judgments still abound with myth and subjective opinion. In a sense, this book is about the process of myth formation. It is a sociology of practical forestry. How do different people arrive at contradictory beliefs? What is the internal logic of these belief systems?

In the process of analyzing other people's beliefs, I have inevitably developed my own set of myths. To me, of course, they don't feel like myths. They simply feel like common sense. But listening to so many other viewpoints has taught me a lesson about dogmatism; I have tried not to be seduced by my own rhetoric. I offer my personal voice only as one among many.

There are surprises herein for loggers and environmentalists alike. I have not followed party lines. To my mind, many of the old arguments are becoming stale. Battles are being fought in which everybody comes out feeling like a loser. I'd like to shake up the old myths, to stimulate new lines of reasoning. So if you're looking for support for your own preconceived position, well, you might or might not find it here. If you're ready for an uncharted journey through the politics of the woods, however, this book should prove worth your while.

Ray Raphael
Whitethorn, California
January, 1981

Part I

Voices in the forest: problems and responses

Chapter 1

Tree mining: the voice of history

WHEN European settlers first touched upon the eastern seaboard of North America, they were confronted by a thick and apparently endless forest. Centuries earlier, Europe itself had been covered with a similar forest; but an increasing population, which required an increasingly high standard of living, had transformed the deep woods of the Old World into farmland. Now, the colonists would attempt another conversion in their new home: razing the mysterious, wild forest to make room for fields and pastures. These people perceived the forest more as an obstacle to progress than as a precious resource.

With trees to spare, the settlers developed a technology that utilized wood for a variety of purposes. Houses, barns, ships, bridges, fences, carriages, furniture, tools, barrels — the appliances of daily life were fashioned from the trees that the settlers wished to clear from the land. Wood was burned in every home for cooking and heating; it was also turned into charcoal for the blacksmiths' forges. With the introduction of the steam engine in the eighteenth century, charcoal became the primary source of energy for industrial production. Throughout the seventeenth, eighteenth, and nineteenth centuries, more wood was consumed as fuel than was made into lumber in the United States.

By the middle of the nineteenth century, most of the desirable farmland east of the Mississippi had been cleared of all trees. During the Civil War the remaining forests — mixed hardwoods situated on hilly terrain or infertile soil — were cut down for charcoal to fire blast furnaces for wartime production. By the time the industrial furnaces finally switched over to coal a few years later, the vast eastern woodlands had been totally consumed. Forests once thought inexhaustible had, in fact, been exhausted.

Forestry's age of innocence had ended: the timber resource could no longer be taken for granted. Lacking the timber for both fuel and lumber, several eastern states enacted primitive conservation laws, which offered bonuses and tax incentives for the planting and nurturing of trees. But the forests could not grow back overnight, and the demand for wood products continued after the first generation of trees had been used up. Lumbering interests did not wait around for the better part of a century for new trees to mature. Instead, they moved on to where the trees still stood tall, to forests that had not yet been cut. During the second half of the nineteenth century, the pines of New England were superseded by bigger and better pines from the great North Woods of Michigan, Minnesota, and Wisconsin. But these, too, proved finite in number. As the century drew to a close, the best of the timber in the North Woods had been cut. Once again, the tillable land was turned into farms; on the areas unclaimed for field or pasture, another crop of trees would be a long time in the making.

Farther west, there were trees still bigger and better than any that had been seen before: the majestic redwoods of California, and the giant Douglas-firs, cedars, spruces, and hemlocks of Oregon and Washington. The forests here were on an altogether different scale. Many of the trees stretched 250 feet into the sky and measured 50 feet around at the base of the trunk. A single tract of forested land could be as large as an entire New England state. Confronted with a logger's paradise, the western pioneers boasted that they had at last discovered an inexhaustible source of timber. "California will for centuries have virgin forests, perhaps to the end of Time," remarked an awe-struck admirer in the 1850s.[1] There was no talk in the Old West about trees being a "renewable resource"; if the virgin timber would last forever, the trees didn't have to be renewed. Timber was extracted from the surface of the earth in much the same way as minerals were mined from underneath the ground. There was no more thought of replacing a 250-foot tree than there was of returning gold dust to the streams.

The unsettled tracts of western timberland were so vast that the government had to develop special strategies for dispensing with them. Some fell into the hands of the railroads as part of the land-grant rights-of-way, but the avowed goal of the federal government was to disperse the ownership as widely as possible. To this end, the famous Homestead Act of 1862 was supplemented by the Timber Culture Act of 1873: a homesteader could add 160 acres to his claim simply by agreeing to grow trees — that is, not to cut them down — on 40 of the extra acres that the government gave him. The Timber and Stone Act of 1878 enabled citizens of the West Coast states to purchase 160 acres of public timberland, for personal use only, at the nominal charge of $2.50 per acre. In this manner the spoils of the ubiquitous timber could be distributed equally to a democratic army of independent loggers, the West Coast equivalent of the small, independent farmers of the East and Midwest.

It didn't work out that way. In practice, the laws were easily manipulated to serve the purposes of more organized logging interests. Companies employed individuals to purchase public timberlands at the nominal fee and then turn over the deed. Homesteaders who took up the 160-acre timber preemption satisfied the government by preserving 40 acres of trees, but then quickly sold off the timber on the remaining 120 acres to whichever company happened to be operating in the area. The trees were thus made available to those who could most readily utilize them: the local logging empires that were emerging up and down the Pacific Coast.

Only in rugged terrain or inaccessible localities did the democratic distribution of timber work as intended. Far from railroads or rivers, homesteaders in the backwoods harvested products of the forest primarily for use by themselves or their close neighbors. Occasionally they processed the trees into small hand-split products that could be transported out of the woods on the back of a mule: shakes, fence posts, railroad ties, grapestakes. But the logs had to be processed right in the middle of the forest. The mules could not carry uncut logs to the mills, so full-scale lumbering remained impractical in the backwoods.

For the most part, the ideal of settled, self-employed loggers could not be realized in the West. Nothing short of a fully organized crew working in a coordinated effort would suffice to move the giant timber out of the woods. In the eastern and Lake states, a homesteader had only to wait for winter if he wanted to log by himself: he then placed his logs on sleds and had his horses slide them out of the woods over the snow and ice. In the West, though, the trees were too large and the land too uneven for such a one-man operation, and there often was no snow or ice in the winter. Here, logging required a new assortment of techniques utilizing a unique combination of gravity, water, animal power — and human sweat.

BIG TREE TECHNOLOGY

The first problem faced by a western lumberjack, strangely enough, was to climb the tree to its trunk. At ground level the swell in a giant redwood or cedar might be eighteen or twenty feet in diameter, while ten feet up the trunk would shrivel to a mere twelve or fourteen feet. To save work and avoid the pitch in the stump, the fallers climbed the tree by chopping a small hole through the bark and inserting a springboard, a flexible plank. Often, three or four springboards would be used as a ladder to

reach the desired height. Working from these portable scaffolds, a pair of fallers could chop down the tree with axes, a task that might take a day or two for each tree.

Once the tree lay safely on the ground, the buckers removed the limbs and cut the trunk into logs. With hand-jacks and peavies, the lumberjacks rolled the logs around stumps and down the sidehills into the gullies. There, teams of oxen or horses would drag the logs to the main skid roads. Since there was no snow or ice upon which the logs might slide, roads were constructed of solid rows of timbers, which had to be watered or greased to minimize friction. The bull-puncher or bull-whacker (his name varied with the location) drove dogs (hooks) into the end of each log, connected the logs to his team with a rope, and drove the team down the skid road to the nearest mill or river.

Skid roads were seldom more than a mile or two in length, for overland travel was difficult and tedious. In the early days, the preferred means of transportation over long distances was by water. The presence of any type of water — oceans, rivers, and

The ox team

even streams — facilitated the movement of logs. Along the Pacific Coast, logging schooners anchored in "dog-hole" ports to take on their cargo. Sometimes primitive wharves were constructed; more often, the logs were loaded by cables suspended from the nearby bluffs. Inland, the logs were dumped into the still backwaters of the rivers during the dry summer months. If several outfits operated along the same river, logs were branded with the insignias of the owners and sorted out down at the mills. When the high water came at the end of the season, the logs tumbled downstream. If a log got hung up on a shallow riffle, the "river rats" pried it loose with their peavies. If all else failed, the troublesome log was hitched to an animal team on shore, which dragged it over the riffle to deep water.

In most areas, it was only a matter of a decade or so before the forests immediately adjacent to the rivers had been exhausted. To reach farther into the hinterlands, the loggers had to utilize smaller and smaller streams. In order to trap enough water in these tributaries to transport the logs, the loggers built splash dams upstream from the decked timber. When released, these dams created man-made floods that literally flushed the logs down to the mills.

Soon even the timber along the streambanks was stripped away. Millions of acres of untouched forest remained throughout the West, but the trees could not be reached by water. Full-scale harvesting in the backcountry awaited the development of a mechanized overland technology that could transport giant logs over extended stretches of rugged terrain.

SID STIERS

LUMBERJACK AND SAW FILER

"My grandparents come out here in the fifties. On my mother's side, eighteen and fifty. My dad's folks come in about eighteen and fifty-six or seven. Worked in the woods — that was all there was in them days.

"They had ox teams at about that time, and then it wasn't long till they got horses. Team of good horses will outpull a team of good oxen. Faster. Them old oxen was slow, you know. They'd just mope along there. But when they come out here, you see, when my grandparents come across the plains, that's all they had was oxen. The horses was zero. In the East they had a few horses, but out West was empty country. Finally they got to bringing horses in, see.

"I was seven years old. My dad was logging up here on our old homestead. Had quite a lot of timber there. He logged it to go on the river drive. I run the old water sled. I was big enough. I didn't have to know much, the old horse knew more than I did. Just had a barrel on a sled, and these skids was so far apart. Two runners on it, planks across it, and a big old fifty-gallon barrel set on that. You'd get her started out ahead of a string of logs — gosh, I don't know how many logs. After they got them on the skid road, they'd trail them. So I'd go along with that and pull the pin in the barrel, and that water'd string out along there and help them skid. A lot of them used grease — go along with a bucket of grease and a mop ahead of a big turn of logs. Those logs, if you didn't have the skids watered or greased or something like that, they'd just pull harder than heck. Made worlds of difference.

"The old river drives were about turn of the century, along in there. All the men would dump the logs into the river till they got their logging done. They'd get together, all of these outfits that was dumping in the river, and they'd get a big log drive a-goin'. Took quite a time. They'd start in the fall. Maybe it'd rain a little and raise the water, and then they'd take off. They'd end up way late at the old millrace down there at Eugene, where it takes out of the river. Big logjams. They'd have to pull the logs out to get them going. Farther down the river, all the logs would get hung up on a riffle. Just one riffle after another, they'd have to pull the logs off.

"They'd get the logs into the millrace, and it'd be freezing. They had to wade out and drive the dogs in the logs, and then they had a guy who rode the logs out. They'd get on a log where it's about waist deep and off they'd go, hollering all the way. Cold! But once they got wet it wasn't so bad. They'd get wet anyway, so a lot of them would just jump in and get it over with.

"One time they was logging sugar pine way up the Willamette. So they logged that sugar pine and put it in the river. They even built a dam. The Willamette up there isn't too big, so they built a flood dam. They got a few logs in and they turned the water on and by-golly: nothing! The logs just laid there. They'd sink, see, just go down like a rock. Sugar pine don't float. I don't know why they didn't know they'd sink. They had to go down there with hooks and fish them all out.

"I was too young to work on the river drives. I was just a pest along about then. Later ones, I was probably about twelve. First job I really went out on, I was sixteen years old. I was bucking wood for an old Willamette swing donkey. I worked in the woods, on the rigging or anything, till nineteen and seventeen when I got a job filing the old hand briars, the misery whips. From then on I did some falling and bucking, you know, but mostly my job was filing old hand briars. That was my trade for about thirty-five years.

"Worked for lots of different outfits. Lived in logging camps. Dirt floors and straw mattresses. It was a mess, you look at it now, but we didn't think much about it in them days. One place, they had just a little old rake and they'd rake it out every once in a

Peavies poised, turn-of-the-century Washington
"river rats" release logs back into the current.

while, when you couldn't get in. Those old beds were just made out of boards: poles up and a bottom on it, and you'd fill that with straw. When it got too hard, you throwed that one out and built yourself another bed of hay. Had to carry your own bedroll. They'd have a row of bunks clear around the wall, two bunks high, one at the bottom and one at the top. And benches to set on, and that's all they had. Set there to put your cork shoes on.

"The old cookhouses, they fed good. But, boy, you get a big crew sitting down to the table, and you had to be fast or you wouldn't get much to eat. Fellow had to raise up and reach clear over the table, if he wanted to get a good hunk of meat. By golly, it was quite a deal. They had good cooks; mostly women cooked in those camps. Just common meat, potatoes, and vegetables when they got them, and they'd bake pies and stuff like that. Lot better than you get now. That come from the ground up, what they cooked. If they didn't put out the food, well: 'Ain't nothing to eat here, I'm going someplace else.' And away they'd go.

"It was rough living in them camps. West Fir was one of them in the twenties. To get a place to live I lived in a tent. All I had was a little, wood, two-door cookstove. Tough going. If you wanted to work, you had to put up with it until something come along better.

"Booth-Kelly, they was a bigger outfit. They had these houses built on skids. So they'd move them on a flatcar, and then when they got to where they was going to have a camp, they'd slide the houses off. When they got to working quite a ways off in some other direction, why they'd slide the houses back on and move to another camp.

"You know, there was always a joker around one of them camps. Worse then than it ever is now. This one guy he got up early and built this fire up good and hot and then wet on it and away he went. Man, I tell you the covers was a-flyin'. But they got even with him somehow. Son-of-a-gun, there was always someone pulling something on somebody."

Pranks may have occurred, but the logging camps were not noted for high living. In most camps, rowdy behavior was cause enough to be fired, as was socializing with the cooks or waitresses.

The days were taken by work, and at night silence was enforced in the bunkhouses, in order that the men might sleep. In some camps, a strict silence was even enforced at mealtime.

The inhibitions of camp life, however, were relieved by periodic trips to the nearest city or town. Because of the seasonal nature of employment, nearly every lumberjack spent a good part of the year out of the woods. Near the mills on the outskirts of town, shanties, cheap hotels, saloons, and restaurants sprang up to care for the transient workers. These sections became known as "skid rows," named after the old skid roads upon which they were located. Life on skid row was the reverse image of life in a logging camp: there was no work to be had, no restrictions on personal behavior, but plenty of vices upon which money might be spent.

Seasonal migrations of the workers were paralleled by the frequent relocation of the logging camps themselves. As each new section of the woods was logged out, the center of operations would move on. Such an unstable situation was not conducive to a settled life among the loggers. More common than the family man was the "boomer," or tramp logger, who followed a loosely plotted trail from camp to camp. According to Rex Seits of Lacomb, Oregon: "The old tramp logger used to, well, he liked to see the other side of the hill once in a while and learn new games, you know, different ways. That's their life: just to keep moving; although they're just out hunting for a good cookhouse. And your companies at one time would just as soon you did, because when a new man come in, why, they'd get a lot more work out of him. He was faster, quicker, energetic; he was always in a hurry. After he was there a while, maybe he'd slow down."[2] Jess Larison, a timber faller from Dexter, recalls: "I spent several years in those camps, and I never saw a single one of those men again."

The mark of the tramp logger was a bedroll on his back; in self-mockery, he called himself an "A.P.A." (American Pack Animal).[3] As often as not, his bedroll was infested with fleas or lice, which were easily spread from bunkhouse to bunkhouse. When the radical Industrial Workers of the World organized in the woods just before and after World War I, their rallying cry struck home in the hearts of the lumberjacks: clean bedding should be supplied by the companies at all logging camps. The call for clean bedding probably won more converts for the Wobblies than did the demand for higher wages.

PICKING UP STEAM

By the time the workers in the woods became organized, mechanization had altered the fundamental nature of their occupation. Two types of steam engines — railroad locomotives and high-powered winches called steam donkeys — had revolutionized the transportation of logs. Railroad tracks could be laid where water never flowed, while steam donkeys could pull logs both faster and easier than the cumbersome animal teams. The steam era ushered in the famous "highball" logging shows, in which production and efficiency surpassed all previous hopes and expectations. Unlike the old river drives, the railroad operated year-round; unlike the old animal teams, the steam donkeys were mindless machines that had neither independent wills nor stomachs to feed. Indeed, the only "food" required by the railroad locomotive and steam donkey was wood — and there was plenty of that around.

The first steam donkey was put into use in 1881 in northwestern California by John Dolbeer, who took out a patent on his invention two years later. The design was simple: a steam-powered engine reeled a cable onto a spool, with a log attached to the end of the line. The engine, anchored to a sled, could pull itself along a skid road with its own power by use of a block and tackle hung on a nearby tree or stump. Logs could therefore be yarded from any position along the skid road.

The early donkey engines worked side by side with animals. The steam donkeys yarded the logs from the woods to the skid roads, where the animals took over. After each load was yarded, a horse — or a man — had to drag the cable back into the woods so it could be secured to another log. Gradually, however, the techniques of donkey-cable logging were streamlined and perfected. During the 1890s, a more powerful version called a bull donkey began to replace the animal teams on the skid roads. With the addition of a separate spool to the engine, an extra haulback line was able to lead the mainline back into the woods after each turn was completed, eliminating the need for men or animals to carry around the cable.

In the early systems, the logs were dragged along the ground where they encountered excessive friction, ran into obstacles, and carved deep trenches in the earth. By hanging the mainline from a stump, however, the front end of the log could be lifted slightly off the ground. Through the years, stump rigs climbed higher and higher, until they were 200 feet above the ground on specially prepared spar trees. From a single spar tree on a landing, a high-lead system could stretch out in a fan-shaped configuration and log everything within a radius as long as the mainline. In the most sophisticated forms of cable logging, both ends of a line were suspended from trees, enabling the entire log to be lifted off the ground as it was yarded. (These skyline systems are enjoying a renaissance today because they minimize soil disturbance.)

The impact of the steam engine on the woods was profound: more trees could be cut from more distant locations. Terminal logging railroads could be built in the vicinity of any prospective mill site to provide major arteries into the forest. When operations were finished for a given site, both the mill and the tracks would be taken up and reassembled amidst an unharvested tract of land. Reaching out from these short, specialized railroad lines, the bull donkeys provided feeder systems to the right and left, and these in turn were fed by the yarding engines, which extended their cables directly to the logs on the ground. Since yarding with steam donkeys was twice as efficient as with animal methods, once-marginal lands could now be profitably logged.

Clearcutting became a practical necessity in donkey logging, for the cables needed room to maneuver without running into standing trees. The railroads, too, led indirectly to clearcutting: the capital costs of setting up shop at any particular location required the company to take all it could get before moving on. No longer were only the best and most accessible trees taken; anything that could pay its way out of the woods wound up at the mill. As for trees that could not pay their way, they were either burned on the spot as impediments or burned as fuel for the donkeys and locomotives. Indeed, the railroads themselves became significant consumers of timber for ties, tunnels, and trestles.

As the extent and scope of the cutting increased, forward-looking citizens began to fear that the expansive forests of the West would follow the fate of the eastern woodlands and soon become exhausted. In the words of Gifford Pinchot, Forest Service director in the early 1900s, "The United States has already crossed the verge of a timber famine so severe that the blighting effects will be felt in every household in the land."[4] As a direct consequence of accelerated logging, a strong conservation movement developed around the turn of the century to push for the protection of timber resources. But "protection" at that time did not entail preservation as wilderness; the conservationists only wished to ensure that there would be trees to cut down in the future. In 1897 the Organic Administration Act set the goal for the newly formed National Forests: "to furnish a continuous supply of timber for the use and necessities of citizens of the United States."[5]

The interest in sustained-yield forestry was not limited to public lands. In 1908 Teddy Roosevelt convened a White House conference that included governors, members of the Cabinet and Supreme Court, scientists, and active conservationists. The conference adopted a "Declaration of Principles":

> We urge the continuation and extension of forest policies adapted to secure the husbanding and renewal of our diminishing timber supply, the prevention of soil erosion, the protection of the headwaters, the maintenance of the purity and navigability of our streams. We recognize that the private ownership of forest land entails responsibilities in the interests of all the people, and we favor the enactment of laws looking to the protection and replacement of privately owned forests.[6]

These principles, however, were never backed up by federal legislation, and a third of a century passed before any of the western states took such concerns seriously. The writing was on the wall, but for the time being the old-growth trees were still too numerous to count.

LIFE AND DEATH IN THE HIGHBALL CAMPS

The effects of steam power were also felt on social and economic levels. Because of the high capital costs in railroad-donkey logging, small operations found themselves struggling under a distinct disadvantage. Logging concerns lacking the money to finance railroad tracks and steam donkeys were driven out of business by their aggressive competitors. The mills, meanwhile, were modernizing with double and triple circular saws, band saws, edgers, planers, and a hefty assortment of log-moving machinery. The larger outfits were better equipped to manufacture a wide variety of products, just as they were better equipped to extend their railroads and donkeys deeper and deeper into the woods. Indeed, the most successful companies were vertically integrated to include not only the extraction and manufacture of the lumber, but also its transportation and sale in distant markets. The big companies owned the timber, the logging railroads, the mills, the lumber schooners that cruised the Pacific Coast, and the distribution yards in major population centers such as San Francisco.

Back in the woods, the logging crews were larger than ever. Cable logging was a complex affair, requiring precise coordination among the separate tasks. When the choker setters had attached a log to a cable, the whistle punk signaled the donkey puncher that it was time to reel the line in. Once the log was on the landing, the chaser unhooked it; the loaders and railroad men took over from there. For each logging show, there was a variety of specialized personnel: hooktender (woods boss), saw filer, donkey doctor (mechanic), and a full-time bucker just to supply the fuel for the steam donkeys. With the camps now housing a minimum of 50 men — sometimes up to 200 — a host of service personnel was also required: storekeepers, time-punchers, bunkhouse lackeys, and, of course, the cooks. The companies became wholesale suppliers of food, clothing, and other amenities. The same railroads that carried the logs out of the woods returned bearing the necessities of everyday living.

There were more men on the job — and more machines. With cables and logs flying through the air, split-second timing and communication were paramount. Yet speed and production, not the safety of the workers, were the prime concerns of the highball shows; many a highball logger found himself in the path of a log or a swiftly moving cable.

Death was an ever-present reality for the workers in the woods.

In the words of Walt Hallowell, who once worked on a record-breaking crew which yarded and loaded 1,690,000 feet of timber in an eight-hour day: "I've heard of a man getting killed in the morning and they'd put him behind a stump until that evening to go into camp. Now that's what they call a highball camp."[7]

It was no accident that the increased influence of the I.W.W. and other radical union organizations in the woods coincided with the climax of the highball shows. The workers were subordinated to the dictates of productive efficiency. The individual lumberjack felt — and was — of little importance. All large shows had several "utility men" waiting in the wings to fill in for a sick or injured worker, because the interdependent tasks required that each and every position be filled at all times. Jess Larison recalls: "Several times I've taken a man's place where they've just been killed on that job, and stepped right in and did his job. It don't make you feel too goddamn healthy or secure."[8]

Of all the jobs involved in donkey logging, none was so dangerous — or so dramatic — as high-climbing. Once a spar tree had been selected to support the elaborate network of cables, a man had to ascend it with spurs and a rope to cut off the top. It took an adventurous spirit to work in the woods in those days, and high-climbers tended to be the most adventurous of all.

ERNEST ROHL
HIGH-CLIMBER, DONKEY-PUNCHER, AND TIMBER FALLER

"I came by way of Cape Horn. Cape Horn to Callao, Peru, and from Peru we went up to Tacoma to take a load of lumber to go back. But I jumped ship there. I wanted to see another country.

"I started to sea when I was fifteen years old. Had my first drink of whiskey when I was fifteen, going around Cape Horn. The cook can't cook, the crew got to have something, and everybody got a tin cup full of rum. And hardtack, ship biscuits. The cook can't cook, because as soon as he gets a fire started, a wave comes over, the fire's out. But you take a fifteen-year-old boy with a tin cup full of rum, and he don't give a damn if the ship go down and never come back up. That was 1908.

"In 1915 I came to Eureka from Callao. I was second mate.

Then the immigration commissioner advised me to quit the sea, because I was not an American citizen. So I was looking for work. Work's hard to get. So I walked. I walked up where the Redwoods College is now. I'm a sailor, right? What the hell does a sailor know on a dairy ranch? Walked through Fortuna, and I come to Scotia. Up to the office, a guy says to go to the mill and ask the foreman. I didn't know the mill from nothing, but I went into this building. It was a big building. And pretty soon somebody turned all hell loose in there — all the damnedest racket! Must have been right after lunchtime, I don't know. All the machinery, everything, was squealing. Somebody come in and asked me what I was looking for. I told him, 'I'm looking for the place to get the hell out of here. Nothing doing!'

"I took out of there, up the county road to Holmes Flat. There was a new camp up there. Section foreman asked me if I had cork shoes. I could talk good English, but 'cork' and 'calk' shoes I didn't know. 'Oh,' he says, 'never mind.' Takes me to this tall fellow over there, says, 'Put the boy to work, he needs work.' So I got a job bucking wood on a pile driver.

"Then came first rain in the fall: that's it, no more, the woods shut down. So I came back to Eureka. I ran across an Englishman I sailed with on one of the British ships. I said, 'What are you doing, Bill?' He said he was 'on the beach.' When a sailor is 'on the beach,' he's broke. He doesn't say, 'I'm on the bum.' He's on the beach, like driftwood. So I paid his bill — owed a month board and room — and we had a drink of beer. Next day, we walked up the street and he said, 'Ernie, there's a schooner looking for a crew. Bound for Sydney, Australia.' I said, 'Bill, are you all right? A limey and a sauerkraut — going to Australia?' That was World War I. So he says, 'Oh, shank your bloody nationality.' So I shanked my nationality from a German to a Hollander and I signed on.

"Next year I landed back in Eureka. I ran across the pile driver boss and went back to the Pacific Lumber Company. Worked on the pile driver there for a couple of months, then that was all the driving. Had all the trestles done. I got on the spool donkey, pulled rigging. Little side-spooler. Single-engine, just one engine on one side. Pulled the logs in with a cable. You pulled the line out on the back, and the donkey pulled it back in with the log. Used eleven-sixteenths-inch line, eight strand.

"On the nineteenth of June, I had an accident — another young fellow and I. Rigging let go of the stump. They didn't use straps, they used grabs, chain grabs. There was a hook on one end, then about three foot of chain. Put that hook in the stump, then you hook your pulley in the end of the block, and then your rigging. Well, one of them chains pulled out of the stump and let go and killed the guy standing alongside of me. Cable knocked me forty feet, killed the other guy. Young fellow from Arcata. The only thing that killed that fellow, he had a habit of either sitting down or leaning up against something. Him and I was side by side, only I was standing free and he was leaning up against a windfall stump, and he took the solid blow. There was no give, he got the full force. I was loose.

"Well, I lost couple, three days, then went back on the rigging. Then built skid roads. Early days, they had crosstie skids. In 1916 they started in putting them long poles, fore and aft. Some of those logs were seventy, eighty feet long; you just stuck them in. If you could get just forty, use forty. That's how you build your road.

"You put your yarder ahead. They got them eleven by thirteen donkeys, some places got bigger donkeys. They pulled those logs to the skid road. The bull donkey is down on the landing. He pulls it down the skid road to the landing. On the landing, load them onto the train.

"The bull donkey goes on the length of his line, 2,400 feet. All right, the yarder is ahead of that, logging to the skid road. Then the water-slinger, he goes up with the bull donkey, ties onto his logs. He builds his own load. Head log got to be a big butt log. They have all the way up to twenty, twenty-eight logs to the load — that depends on the size of the logs. Then that bull donkey will take those logs down to the landing, pull them in. Then he goes back after another one. When that section is loaded, then the yarder moves ahead and the skid road is built ahead another couple thousand feet.

"Old days when they had skid roads, there was no such thing as take your donkey and put it in the car and take it out there and load it. You had to move your donkey overland. Sometimes it took you two, three days to get you from one setting to another. Moved by its own power. You strung out the line, using blocks and pulleys; it pulled itself up.

"Nineteen eighteen the Pacific Lumber Company started high-lead logging. First was ground logging — all your logs drag on the ground on the skid road. But then 1918 they started high-lead. So they had a fellow there to raise the first pole. All the guy lines had to tighten the same time — that's how it worked. First fellow broke the guy line and down he come. Then the camp boss tried to raise that spar tree. He was in a hell of a hurry, and he broke it. Then the superintendent said, 'Let that young sailor raise the pole.' I said all right, so I went over and raised the first spar tree — it wasn't only but about 120 feet. We got that rig up, put the block up, and the mainline. Pretty soon, they broke a mainline. Look who gets up — that sailor. The sailor had to go back up and put the lines through. Every time they broke the lines, sailor had to come.

"I was boss on one of them little spool donkeys. So one day I asked the superintendent, 'Say, how's the chance to get fifty cents more a day?' — because I had to go up the pole every time the line broke. He said, 'No, if I raised your salary the board of directors would fire me.' The next morning the little sailor was gone. I went up in the hills. Stayed there for a week. Come back down, the superintendent wants to see me. I started back for the hills, and he come out from behind the stump, him and the head time-keeper from the main office. 'Ernie,' he says, 'we raised your wages from four and a half a day to two hundred and twenty-five dollars a month. But we don't want you to quit every time a fellow drops his hat.' I say, 'You tell them guys not to drop their hat, and I won't quit.'

"Then I start topping the trees. Redwood. They would like to have the high-lead block 200 feet off the ground. So you always had to top your trees at 210, 220. You shimmy up, top your tree, put your block up, and put up your rigging. That's your spar tree on the landing. Felt just like the mast out on the old sailing ship at midnight. You don't see no landing, no nothing. I could go up the tree fine.

"Then in '27 they started slackers. You had two spar trees, strung a two-inch cable between those trees, about 2,000 feet. They slacked the mainline, they had a carriage on there, and you hooked the chokers onto the logs and hauled them on to the landing.

"Nineteen twenty-eight the first cats come in the woods. Finally wound up it's all cats. The cats, they do a lot of damage to the ground all right. But early day logging was easy on the land.

"Woods accidents, they used to have some bad ones in the early days. Lots of fellows got their legs broke, pushed between logs. I had this leg broke in '28 and busted the other one in '29. In the old days, a lot of the accidents was cables. Either the rigging pulled out of the stump, or the strap broke, or something. Or logs. Some got the logs rolled over, pinched in the logs. Quite a few got killed with widow-makers, limbs. Sometimes a loose branch hangs up there, and at the least little movement down it comes and hits you.

"One time we were working on a family tree. We had a saw in there, and pretty soon it stuck. We had to beat it off by hand, chop with an ax. So Ernie [Ernest's son] started in, and I started in on the undercut. Pretty soon that thing starts wiggling. 'Ernie,' I say, 'you get the hell out of here.' It was a family tree, a group together, all tangled up. He says, 'You don't know what's coming.' But I say, 'By God, I do.' I gave it two more licks and the tree popped. I took off. Ernie hollered, and it's a good thing he did: 'Dad, look up!' I just froze right there and I looked up and a chunk hit me in the kisser. If he didn't holler, I would have had it right between the ears. Got my eyes, bashed in my nose, everything. Knocked me down on the ground. My hat was gone, and whenever I'm without my hat I'm lost. I say, 'Ernie, Jesus Christ, get me my hat.' The eyeball it was popping around there someplace, popped out. Then I had to walk back up the hill.

"The worst mess I seen was a fellow and his partner had a forked tree. Put in a couple of sticks of powder to blow 'em apart. Tree didn't fall when the shot went off. Fellow started back down toward the tree. All at once it started to pop. Henry went one way, partner went the other way. That would have been fine, but he changed his mind. He doubled back, and just when he came out from behind, it came down on him. That man, if I didn't know him I couldn't have swore it was him. The only thing was his scalp. Flattened him out. One arm lay forty feet from him.

"One railroad accident I know of. Man of German descent, Fritz Speck, a superintendent from the Pacific Lumber Company. They had a derailment. In the back end they had a car with a bunch of hogs in it. Around this old-time camp, they had a cookhouse that fed 200, 250 men. They'd take pigs up there, fatten the pigs. This train had a hog car, and in back of that they had a caboose. So Fritz Speck, he was there in the caboose and broke his leg in the derailment. The brakeman says, 'Mr. Speck, how are you? You hurt? You hurt?' 'Never mind me,' Speck says. 'You watch the hogs. Don't you let *them* get away. Never mind me.' He had a broken leg, but he didn't give a damn about that as long as the hogs stayed there. I admired that fellow."

INTERNAL COMBUSTION: TREE MINING COMES OF AGE

The railroad days in the woods are over. Gasoline power replaced steam power in the second quarter of the twentieth century, making a dramatic impact on both the woods and the men who worked there. The internal combustion engine came in three distinct forms: the chain saw, the logging truck, and the caterpillar tractor. Chain saws replaced the old misery whips, enabling every modern-day logger to outcut even the best of the old-fashioned lumberjacks. Logging trucks replaced railroad cars, eliminating the need to lay down and take up tracks every time a logging show had to switch locations. Caterpillar tractors replaced the old steam donkeys, eliminating the need to construct an elaborate cable network for each neck of the woods. The same caterpillar that yarded the logs could also build the roads. Far more mobile than its steam-powered predecessor, cat/truck logging overcame previous topographical and geographical limitations. The cat/truck combination was the final answer to the problem of accessibility: there is now no area too remote for a viable logging operation. The harvesting of timber can — and does — occur anywhere and everywhere, not just a mile or two from the nearest river or railroad.

Because the trees could now be cut more quickly and in a

greater variety of locations, gasoline power hastened the depletion of old-growth timber stands. Timber companies and gyppo
loggers accelerated their cutting rates on private land as new
roads opened up rugged terrain in the backwoods. As of the
1940s, the market had no trouble absorbing the extra production: World War II and the postwar housing boom created an
unprecedented demand for lumber. Cutting proceeded apace for
about twenty years, but by the 1960s the effect of stepped-up production was becoming apparent: privately owned, old-growth
timber was on the verge of extinction.

Logging comes of age

In order to maintain a supply to meet the demand, loggers had to shift their harvesting operations to public lands, which had been removed from production by the conservation movement in the early 1900s. In the 1930s, more than twice as much timber was cut on private land as on public land in the Northwest; by the early 1960s, private timber accounted for less than half the cut.[9] As the burden of logging in the West shifted to government lands, mill owners and lumberjacks found they no longer could determine the fate of the woods by themselves. Forestry had become a matter of public policy. Although new tools had given the loggers more power to overcome physical obstacles, a combination of dwindling resources and public pressure began to diminish that power.

The increased harvests generated by gasoline power spurred a second wave of interest in forest conservation. When logging trucks and caterpillars started their penetration of the backwoods in the early 1930s, the federal government once again took tentative steps toward regulating forest practices. In 1934 the Lumber Code Authority, established by Franklin Roosevelt's National Recovery Act, called for regional conservation rules that would prevent damage to young trees during logging operations and provide for the replanting of cleared land. A few of these regional regulations were in fact drafted, but they quickly became inoperative when the National Recovery Act was ruled unconstitutional a year later.

Despite this setback, the Roosevelt administration showed a continuing interest in regulating logging activities and, perhaps, even nationalizing privately owned timberland. The timber industry, wary of federal restrictions and afraid of nationalization, began to push for state legislation that would be less threatening and easier to control. Ironically, the industry itself helped draft the early forest practice acts in the West. In 1941, Oregon passed a law requiring loggers to leave 5% of the original stand as seed trees; in the pine country in the eastern part of the state, the law required that all trees less than sixteen inches in diameter remain standing. In 1943 California passed its own minimum diameter law, which prohibited the commercial cutting of coniferous trees less than eighteen inches in diameter. In 1945 Washington passed a law similar to that of Oregon, and California developed a unique Forest Practice Act, in which all rules were to be voted on and approved by property owners representing two-thirds or more of the state's timberland. The rules adopted in this democratic manner were not exceptionally forceful, for the landowners showed little interest in restricting their own activities too severely. The California law lacked any enforcement provisions and was administered by a Board of Forestry dominated by industry representatives.

The impact of these regulations was minimal — and in some cases even harmful. The loggers did not mind leaving a small percentage of trees behind; often, they would have left some trees anyway simply because they were of poor quality. A few conscientious loggers left top-quality trees, but many preferred to take the good ones and leave the bad. This practice of "high-grading," as it is called, amounted to a reverse genetic selection: only the inferior trees were allowed to reproduce. Regeneration from seed trees ran into other problems: by the time a good seed-bearing year came along, unwanted brush and noncommercial hardwoods had often taken over the land. The primitive techniques for regeneration, as reflected in these early forest practice laws, continued to lag behind the advanced techniques for harvesting.

The impact of the accelerated harvest was felt by the loggers as well as the forests. Inevitably, the local booms in the backwoods were followed by busts. Employment ran high for a time, but then the work was gone. With cats and chain saws at their command, the productive output of each logger increased, but the number of men employed in each logging show was correspondingly lower.

Many of those lumberjacks who maintained steady work became commuters. With roads to and from the logging shows, men could live at home and drive to their place of work. With no need to live in lumber camps, it was easier for a logger to be a family man. Even if work could only be found far from home, the whole family might pack up and join the logger in a house trailer. For the logger whose place of work must change year by year, trailers and mobile homes have provided a compromise between a transient and a settled way of life.

Loading the truck: Oregon, 1942

On another level, the takeover of the woods by machinery has de-romanticized logging. To fall and move a ten-ton tree is no longer the ultimate challenge for manly strength. The ancient cry of "timber" cannot now be heard over the motorized hum of chain saws, cats, and trucks. In the words of Bob Ziak, Jr., a life-time logger along the Columbia River: " 'Course now, today, we have the power saws, and they make plenty of noise, but falling timber was comparatively quiet. You'd just hear the swish, swish, swish of the saw and it wasn't even that loud, and then once in a while you'd hear the ring of the wedges as they'd wedge their tree. Today I don't hear the sound or the cry of timber . . . and it was a beautiful sound."[10]

For a pair of old-time timber fallers working by hand with their misery whip, a giant tree was a formidable adversary worthy of great respect: "He's one of us. He's tree. I mean it's you, the next fellow, and the tree. There's three people, I guess you might say."[11] Today, the trees fall quickly, one after another, leaving a man no time to ponder their demise.

Chapter 2

Tree farming: the voice of industry

T HE depletion of old-growth timber has presented the
logging companies with a new challenge: how to re-
create the resources they wish to utilize in the future.
The technological difficulties in logging the giant ever-
greens have been overcome; the problem now is to create tiny
seedlings that can be nurtured into full-fledged trees.

Today, the timber industry considers old-growth forests to be
obsolete. The few virgin forests that remain are valued for more
than just their standing wood fiber; they are seen as productive
sites that can support future crops of trees. Forested land is ap-
praised for its potential, considering the trees that *might* be
grown as well as the trees that are already there. The old "mining" approach to timber was a static one: once the existing resource was depleted, the forest was worthless to the loggers. Now,
the concept of a forest is more dynamic: growth rates are as important as standing inventories. The emphasis has switched from
extraction to production.

Old-growth trees are still cut down, but the logging is supported by a new rationale. The timber industry, in its concern for
production, regards virgin forests as wasteful: the rate of decay in
"over-mature" timber often equals or surpasses the rate of
growth, bringing net productivity to a standstill. In the absence
of logging, the rotting trees are harvested by natural means, such
as wind, fire, insects, and disease. Logging operations, therefore,
serve to salvage dying timber for human needs and reclaim the

land for productive purposes. The old growth is liquidated to make room for young, fast-growing trees.

The ancestral giants of an untamed, unproductive virgin forest are no longer the logger's dream; instead, he prefers the rapid growth and high yields of a scientifically managed tree farm. In the words of a forestry textbook, "Timber resources, untouched by man, present a striking parallel with wild prairies awaiting the plow."[1] Just as wild grasses are removed to open up the land for domesticated crops, so must nature's random assortment of trees be cut down, so man can grow the specific trees he prefers. The comparison of forestry with farming is reflected in the contemporary language of timber production: *plantations* of *crop trees* are protected from insects, disease, and *weed trees,* until they are ready to be *harvested* in cyclical *rotations.* Georgia-Pacific, one of the nation's largest timber companies, boasts, "We are managing a garden — a 4.7 million-acre garden."[2]

If forestry is the equivalent of farming, the timber industry reasons that it makes no sense to set aside areas where the tree

The tree farm: rank and file

crops are not allowed to be harvested. The commercial forester sees the preservation of forests in their natural state as a waste of productive timberland. An advertisement by a timber company in a forestry journal shows a picture of a single ear of corn with a $3.25 price tag.

> Suppose farmers found corn too lovely to pick? Or imagine the rancher so moved by the sight of his wheat field that he cancelled the harvest.
> Sooner or later, all crops are harvested. If man doesn't use the bounty for himself, Mother Nature will claim it. Via insects, disease, rain, high winds or low temperatures.
> It's easier to understand when we talk about food crops. Because we can witness the outcome within a single year. With timber the cycle is a lot longer. But the laws of nature are the same.
> So are the laws of economics. Because whenever large portions of a crop aren't harvested, the price of that crop is going to rise. It's true for corn, apples, peas, and beans, just as it's true for the timber that produces lumber, plywood, particleboard and paper products. . . .
> To be sure, the forest holds enormous beauty. But it also holds great promise. To realize that promise, we must remember that there is a time to sow and a time to reap.[3]

The same theme is expressed more directly by H.D. Bennett, a timber industry executive from the Appalachian region: "We have the directive from God: Have dominion over the earth, replenish it, and subdue it. God has not given us these resources so we can merely watch their ecological changes occur."[4]

The idea of tree farming is not that new. Forests have been farmed in Europe for centuries, for the virgin forests there were depleted long ago. In this country Gifford Pinchot, the famous conservationist of the early 1900s, was fond of using the agricultural model: "Forestry is Tree Farming. . . . The purpose of Forestry, then, is to make the forest produce the largest possible amount of whatever crop or service will be most useful, and keep on producing it for generation after generation of men and trees."[5]

Yet the idea of tree farming did not really take hold until the old-growth timber in the West showed signs of imminent exhaustion. In the early 1940s the price of lumber soared, giving timber interests plenty of available capital — and a sufficient incentive — to invest in reforestation. No longer were there untouched parcels of old-growth timberland that could be purchased on the open market with the profits reaped from the depleted forests. The only real alternative was for the companies to manage the lands they already owned to produce future crops. Thus was the American Tree Farm System born.

BILL HAGENSTEIN

EXECUTIVE VICE-PRESIDENT OF THE INDUSTRIAL

FORESTRY ASSOCIATION

"When I got out of graduate school, I got an offer from the West Coast Lumberman's Association to be their forester in western Washington. I went to work on the sixth day of June, 1941, and six days later Weyerhaeuser announced that they were going to manage a property in Grays Harbor County as a tree farm. Our association picked up the idea that Weyerhaeuser developed there and made a program out of it. In the fall of 1941 we recommended to the National Association that they do something similar for the rest of the country. On the twentieth day of January, 1942, our board of directors certified the first tree farm in the United States. I was in the room when it happened. It was right up the street here in an old building that was then called the Portland Hotel. It's a parking lot now, but I'm going to get a plaque put up there someday in the right place, because there's where the American Tree Farm System started. And I was there when it did.

"Up to that point, there had been many attempts in the United States to practice forestry. Most of them were not very successful for a number of reasons, one of which was that we still had hopelessly inadequate protection against fire, so it was impossible to get the government or private owners to invest money in growing trees on purpose, because there was no assurance that when harvest day rolled around there would be anything to harvest. In the meantime, they may have burned up. One of the pioneer companies in this region, the Long-Bell Lumber Company of Kansas City, came out here in 1920 with the idea that they clearcut this old-growth timber and replant every acre, and they started to do it. They built a nursery and when they started operating in about 1923, they planted the first trees, and every year thereafter they planted the areas they cut. But the one thing they couldn't do — they were ahead of their time — was guarantee that they could prevent those plantations from burning up. And in 1938 the whole goddamn bunch of them did. They had a bad fire in there and burned them all up. It discouraged the hell out of that company from doing it for a long time. It took almost fifteen years to get them interested in it again, because they felt they'd be throwing good money after bad.

"When we certified those first tree farms, the whole idea was to get strong public support for improving our protection against fire, for recognizing that it takes a long time to grow trees. From the time you plant a little two-year-old tree to the time it's ready to harvest, every year you run the uninsurable risk that something could happen to it: it could burn up, it could blow over, it could be killed by insects, it could be stolen by somebody if it's accessible. And you've got to pay taxes every year on the land. Taxes, protection costs, and the accumulation of risk year after year don't add to the value of timber — they only add to its cost. So all during this period, everybody was speculating.

"Those of us hired by industry used the tree farm program as a vehicle to get public support for good protection, for reasonable taxation. At that time there was a drive on in the United States — a political drive by the Roosevelt Administration — to allege that a long-term crop like timber couldn't be handled by anybody except the government. The government would either have to grow timber on its own lands alone, or would have to regulate the private owners. And there was nobody in our industry who looked with favor upon the idea of the federal government coming in and telling us how to do it. So the tree farm program was in part a vehicle to build up some public confidence that here was an industry prepared to do the job of managing these lands.

"My job, and the job of our association all the way through this thing, has been to encourage the people to grow trees, to stimulate their interest by showing them what their opportunities are, and then suggesting to them what they have to do to realize those opportunities. And the record really speaks for itself. For example, in our region we have certified as tree farms more than 60% of all the privately owned forest land in western Oregon and western Washington. To become certified, we require any private landowner — and it could be you or me — to agree to keep the land for the purpose of growing commercial crops of timber. You agree to provide it adequate protection against fire, insects, and disease, or damage by destructive grazing. And you also agree, when you harvest timber, that you harvest it in such a way that you keep the land productive. What that means is that you're going to get it reproduced properly, either artificially or naturally. You can do it either way, but most people today reproduce artificially.

"Originally, a lot of our plans thirty years ago depended heavily on natural reforestation; you cut the timber in such a way that you left the seed sources. But it soon became clear that we couldn't run the risk of letting that land be occupied by brush species. The Old Man Upstairs had built the land into a coniferous forest area; we ought to keep it that way, and not let the hardwoods get it. There's a cycle to the seed crops on the conifers that only averages about two reasonable crops in a decade. If you cut the timber in a period when you have two failures and two light seed crops in a row, you're not apt to get adequate natural reproduction. You get some, but you won't get enough. You end up with a partially stocked stand of timber, and you don't succeed in forestry by having the shelves only partially filled. You've got to fill them up. Better to have too many trees than too few. It's just like your own garden: you want to plant them thick, and if you get too many, you're going to thin them out. You don't want to waste space, otherwise you don't get much to eat. Same way with trees.

"So at the same time we started the tree farm program in 1941, the West Coast Lumberman's Association started a tree nursery. Up to that time, there was no source of trees for the industry. There were a couple of companies like Weyerhaeuser and Long-Bell that had nurseries of their own, and the federal nurseries were not allowed to sell trees to private owners. So we started a nursery in which we got half a dozen companies to agree to buy trees at a predetermined price, and we announced to the world, bragged to the world, that we were going to grow five million trees a year for reforestation. There was no question that part of the idea of getting the industry to put up the money was public relations, but we didn't let it degenerate into that alone. We turned it into a serious program to really get these unstocked lands reforested. And, of course, it's grown like mad. We started out with the desire to grow five million trees a year, and now we grow forty to forty-five million. Being a nonprofit corporation, we sell the trees back to the companies at cost.

"When a tree farmer gets certified, he gets a paper signed by the president of the association, and then he has the right to call on us anytime, free of charge, for any professional assistance that we can give him: reforestation, protection, utilization, anything to do with the management of that property. Mostly nonindustrial owners are certified as tree farms. We spend a lot of money on some of them, but the industry is willing to have us do that, because, to the extent that the landowners outside the industry are doing their job in keeping their land productive, this is going to add to the sum total of timber supply, which ultimately the industry is going to have a chance to buy. That's important to us, because in this industry half of the manufacturing entities in it don't own an acre of land or a stick of timber. They're entirely dependent upon public timber or timber from other private sources. It's only the landowning companies that are anywhere near self-sufficient.

"For years, the big argument for many people *not* to invest money in forestry — particularly for an individual — was that you'd never live to see the results. It took too long. Today if I have twenty, forty, or sixty acres of young trees coming along, and I decide that I need that money to do something — pay medical bills, send a child to college, build a house, anything that requires some capital — why, I'd find a ready market for that timberland. The trees don't have to be mature and ready for harvest. That's changed the picture, that's encouraged a lot of small individuals to go into tree farming. You'd be surprised at the number of businessmen that come to see us each year wanting to go out and get a piece of land and grow trees. I've got a piece myself, a twenty-acre piece up in Washington, and I can say it's the best investment I ever made in my life."

The key to tree farming is intensive management. The complex forest environment is controlled and manipulated to maximize the major element of commercial interest: wood fiber. As Bill Hagenstein puts it, "Forestry has always been an environmental undertaking. It's main thrust has always been taming the wild forest for man's use and enjoyment by managing the ecology, instead of letting it run rampant as though there were no people around."[6] The tools and techniques of modern agricultural science are brought to bear on the woods, tending and caring for the well-being of the crop trees. In the words of Bernard Orell, vice-president of Weyerhaeuser, "This means the application of fertilizers, insecticides, and herbicides, much as we nurture children to the full flower of adulthood by use of medicines, nutrients and preventatives."[7]

The most significant improvement that man can make over nature is to shorten the rotation age for each crop of trees. Nature provides for regeneration, but it does so only slowly. After a fire, wind storm, insect infestation, or other natural means of harvesting the trees, the forest is opened up for fast-growing, sun-loving species of plants. These "pioneers," as they are called, penetrate and enrich the soil, shade the exposed earth, and start to build back the forest biomass by photosynthesizing the sun's energy. Pioneer plants are relatively short-lived and often of limited size; during their period of dominance, however, they serve to create an appropriate environment for the various successor species. Several stages of plant growth may ensue, culminating in a climax forest of mature trees. This process of forest succession sometimes takes centuries to work itself out.

When a mature forest is logged, nature's plan for forest succession is set into action. Timber growers, however, can accelerate this lengthy process with various management techniques. In most forest ecosystems, there are two or more types of trees with commercial value. Foresters can, therefore, ignore the climax species in favor of trees that come along earlier in the cycle; or, they can try to eliminate the pioneer stages through some form of weed control. They plant the desired trees, even if they are successor or climax species, and they do what they can to eradicate all competitors. By imposing a one-step cycle in place of the complex process of forest succession, forest managers can reduce the time required for each crop to half a century, or even less. In the words of Dave Burwell from the Rosboro Lumber Company, "I know it took nature five hundred years to grow that forest, [but] we can do it in fifty, because in this climate the damn stuff grows back faster than we can cut it."[8]

While shortening the rotation, foresters are also simplifying the ecosystem so it is easier to manipulate. Rather than waste space with noncommercial species, they like to stock the forest with only the most desirable trees. And the behavior of the trees is more predictable — and therefore more easily controlled — if they are all of equal age. With even-age trees of the same species, the techniques of intensive management can achieve their most direct results: fertilization and weed control can be applied only when the trees are most in need of assistance, and the trees can be treated for specific diseases at the most susceptible stages in their development.

This sort of even-age monoculture is modeled after standard farming practices. A cornfield is intended to raise corn — and nothing else. The farmer does not plant an occasional pea or bean amidst the rows of corn, nor does he permit the field to be overrun with dandelions. He plants his seeds all at one time so his crop will mature evenly. Come fall, the farmer does not harvest every fifth cornstalk and leave the rest; he systematically "clearcuts" his field.

And so it is with trees, assert some members of the timber industry. Clearcutting is an integral part of even-age monoculture. The surest way to create an even-age stand of a single species is to remove the old forest in its entirety and start with a clean slate. All trees must be cut to the ground, whether or not they are commercially useful. The unwanted residue must be cleared away in order that the soil be exposed for planting. The industrial forester, like the farmer, likes to burn his slash or plow it back into the ground. Corn seeds are not planted amidst weeds and last year's stalks, nor are young trees planted in brush and logging slash. A whole new technology is developing around the use of fire, herbicides, and mechanical equipment to deal with the unwanted material. The science of site preparation is becoming rapidly extended: fire is now used to destroy fungi harmful to crop trees; chemicals are used to remove pathogens from the soil; heavy equipment is used to remove stumps that might harbor root diseases.

With the forest cover removed and the site prepared, the ground is ready to receive its next crop of trees. The new trees, however, will not be like the old ones: they will be faster-growing "super-trees," the genetically engineered products of seed orchards and scientifically managed nurseries.

PHILIP F. HAHN

MANAGER OF FORESTRY RESEARCH, GEORGIA-PACIFIC CORPORATION

"I'd like to talk mostly about the Douglas-fir region. This includes the managed forests of Oregon, Washington, and northern California. Historically, the timber types in this area evolved naturally. Fires would occur, and then nature would seed back the forest. But the seeding of a new forest was not always successful; and if it was, it still took a very long time, because Douglas-fir seedlings had to compete with the other vegetation in the area. Often, some of the fast-growing weeds and tree species would overtop the Douglas-fir, so the area wouldn't come back into full production, unless it was helped somehow. And if it did come back, it would often be spotty and unevenly aged due to the long and hard process.

"Of course, foresters are keen observers: they learn from nature and they soon figure ways to manage the forest better than nature does. Back in the twenties and thirties, the rule was generally to cut out small patches of trees and let the area be seeded back naturally. It really didn't matter how long it took because there was still plenty of old-growth timber available everyplace. As a matter of fact, they should have cut the trees down faster than they did, because the old-growth trees in the virgin forest were deteriorating rapidly. They were hundreds of years old, and often disease got into them because of their low resistance. They had to hurry to get some of those trees out in order to save them. The abundance of timber and low cost of wood didn't permit them to worry too much about the young forest.

"After World War II, the Northwest became more heavily populated, the industry started to build up, and the world demand for forest products increased. This caused increased stumpage prices and more intensive logging. This automatically created a need for a more intensive forest management practice than had been done in the past, and, in turn, ushered in an era where the companies and the government agencies became increasingly concerned about the future growth of the forest, about providing artificial help to renew the forest.

"For example, back in the Tillamook area there were large burned-over areas due to accidental fires, and they did an extensive amount of aerial seeding to bring the land back into production. Helicopters were available at that time, and it was just natural to collect cones during a good cone year to secure seeds and then deposit the seeds by aircraft. Many agencies and companies did a large scale reforestation job this way, mostly during the fifties and sixties.

"But as stumpage increased and, consequently, the value of wood increased, we knew right away that we could afford another way to reforest. Another aspect came into the picture also: a new, worldwide trend to improve our forests through genetics.

"Genetics, of course, is an old technique used in agriculture and animal breeding for centuries. In crops like wheat and corn you get a crop every year, so you can make rapid advances in genetics. If you have a selection program, a crossing program, you can see your results a year later. Animal breeding is a little slower, but even with animals it takes only a few years for a generation. But when it comes to trees, you have to wait for decades before you see actual results from a genetics program. That is probably the reason why genetics in forestry wasn't applied as rapidly, and still isn't applied as easily, as in other areas. But companies and foresters realize that we must get into this field if we want to think of the future.

"In the fifties, Georgia-Pacific and some of the other companies started on these new genetic programs, but we could not produce enough seed in a genetics program to spread the seed from a helicopter. We had to find a new way to reforest our land, and that new way is producing seedlings from a genetically improved source in nurseries. With nursery stock, we can spread the seed over a larger area. In aerial seeding, we use roughly twenty to forty thousand seeds per acre and get maybe a thousand trees, which normally would be poorly distributed over the area. That's wasting seed — and land. But with hand planting, we can spread the same amount of seed over thirty or forty acres quite evenly to cover all usable area. We improve the efficiency of reforestation by not only planting seedlings, but also by spreading the improved seed over a larger area.

"Let me explain how a genetics program works in a practical application. Trees, due to natural development, are adapted to local conditions. This is recognized by foresters, so we use the natural stands in a given area, because we know that's the best stock there is. The trees survived there for centuries; if they evolved there naturally, we really cannot do better. Of course, there is still a lot of variation among trees within those stands; you still see some runts, some average trees, and some outstanding trees. In a genetics program, we are after those very outstanding trees. We choose the very best stands in each local area, and within the very best stands we choose the very best groups of trees, where there is a lot of competition, and out of all that competition we choose the very best individual tree. We call this tree a 'plus' tree.

"We could go out there to these trees and collect cones and seeds, but they are spread all over the area, and they are generally tall, mature trees. We would have to climb up and get the cones each time we had a cone crop. But we don't want to do it that way. Technology provides us with a tool by which we can reproduce those trees identically through a vegetative propagation method. We have to get a part of that tree, and the best part is the top of the tree, because that represents the best cone-producing ability. We go out there and take the top off and take cuttings off the tips of the branches. Those cuttings are then used to reproduce the 'plus' tree through grafting.

"Before we develop our cone-producing orchard, we develop root stock that will be compatible with the grafting material. With a compatible root stock, we are able to reproduce these trees easily and successfully. The grafting helps to bring the trees into almost instant cone production. The cutting that was grafted onto the root stock still thinks it's about sixty years old, and it goes on producing cones even when the grafted tree is only two or three feet tall. In an orchard, this is very important because we try to collect cones while we stand on the ground. We don't like to climb trees if we don't have to. We prune them back, but sooner or later the trees will get away from us because they're fast-growing trees, and then, of course, we'll bring in machinery to help in collecting the cones.

"The orchards are generally located on relatively flat terrain. We can easily plant the trees out there, we can thin the trees, we can irrigate them, we can fertilize them, we can cultivate around the trees, we can use mechanized equipment in taking care of them and for collecting cones.

"There is another advantage to having an orchard. Imagine: you have 300 outstanding parent trees, and they freely mix with each other. They improve upon themselves through natural crossing. We also use controlled pollination. This way, we are able to test and further select to improve even on the best trees. Naturally, we have to test them out carefully. They are all good, but we want to find out which are the very best. The testing is done on an ideal site, which we call a progeny test area. These tests give us a lot of information about the trees. We are interested not merely in how they appear in the forest at selection time, but how they pass on their traits to their offspring. We want to find out which trees grow rapidly in the early stages, because Douglas-fir needs the sunshine and the trees need to grow rapidly, so they can compete with the brush by staying on top. And, of course, we are interested in trees that are disease-resistant. They might enable us to avoid using pesticides. Disease resistance is easy to test for: we inoculate the trees and subject them to disease, and the ones that are not receptive to the disease will survive. And we also look for desirable branch characteristics. And we test for specific gravity, the density of the wood. Trees with higher specific gravity will provide a stronger wood and will produce more pulp, so we'll have a higher return. We test for actual growth rates, height growth, diameter growth. We can test for all these things. It's all done scientifically in a statistically sound system. This way we can sort out the trees that have good wood quality, are disease-resistant, and grow rapidly. Of course, nature always has done a good job of selection, but we expedite the selection ourselves.

"In order to do a large reforestation job, we have to have a proper nursery facility. In the past, standard practice was to produce bare-root seedlings, which were grown out in an open field for a couple of years and, when they were ready, were lifted, packaged, put in a bag, and planted. That was a well-developed system, but we are in a new era now, and technology improves, and we soon found that there are other ways to produce seedlings in large quantities. So we got involved in containerization.

"The container technology was still in its infancy when we started developing our nursery facilities in 1970. There was no equipment available, there was no proven greenhouse system. I knew the trees needed some artificial help, but I knew I didn't want to baby them. They have to go out in the field, and they have to stand up under very adverse conditions. So I designed a greenhouse system that works like a convertible car: when it's cold, you close it; when it's warm or hot, you can open the sides and the roof. We even put a shade screen over it, so the trees will feel more comfortable, but will still be exposed to a breeze moving through the greenhouse to keep the disease problem down.

"To produce several million seedlings sowing by hand would take a long time. So we had to design equipment to do this. We invented the seeding equipment, so we can set up an assembly line to move the blocks of containers through rapidly. With our sowing system, we are able to seed about 300,000 cavities a day while using ten people. This is considered a fairly good production rate. This way we can get the job done in about six weeks, even if we have to sow about ten million trees, as we do right now.

"When we started out, we were skeptical about how to plant out in the field. We went back to the drawing board. I designed a backpack to carry the blocks (styrofoam containers); this pack can eject just one quarter-section of the block at a time. Then a tree planter can place the quarter-block in a belt holder in front of him and have direct access to the trees. He can just pull on the seedling, and the whole plug of dirt will easily slide out, because the container cavity is tapered. With a dibble he can make an impression in the ground the same form and size as the container plug, and then just place the seedling plug in the soil cavity on the cutover land without disturbing the root system. The tree often does not know the difference, whether it is sitting on a nursery bench or is out in the field.

"There are a lot of angles to containerization. In an outdoor nursery, germination takes a long time, some seeds germinate rapidly and others don't, and you get an uneven crop. In a container nursery, they all germinate fast, because of the artificial help, and you get an even crop. They're off to a good start, and you can gear the watering program and the fertilization program to evenly germinated and evenly spaced trees. Since they have equal growing space, they can grow rapidly with no interference from the neighboring trees. If we sow in the beginning of April, we won't have any difficulty in reaching the maximum height by the middle of July. Then we taper off with our growing and start our hardening program. These trees have to go through a hardening phase, because they have to be conditioned to go to the field. This way we can have them ready for planting in the winter, when the trees are dormant. We can grow one crop in one season, and that's important so we can utilize our nursery facilities well and are able to schedule our reforestation

Test-tube timber: a Douglas-fir tree plug

better. We don't have to wait two or three years to get a crop.

"Before we actually plant the seedlings in the forest, we have to clear the site. A successful reforestation effort actually starts with the logging. After harvest, the limbs and the tops of the trees and some of the other brush stay behind in the form of slash. This can cause a problem in getting onto the land physically. So, it's common practice to burn some of this material to make the area accessible to the planters. This also exposes the ground. Douglas-fir likes an exposed soil. This is known through experience; the companies are not just burning for the sake of burning. We have to prepare the site, and, of course, nature often prepared the site by burning, which resulted in outstanding crops.

"Often, you have too many brush species in the forest. In a case like this, you have to go in and use mechanized equipment to actually remove this material; either cut the stems down, or push them down, pile them, and get rid of them. We really don't like to do this if we can avoid it, because it's expensive; but, on the other hand, it does open up the ground for cultivation work. It's

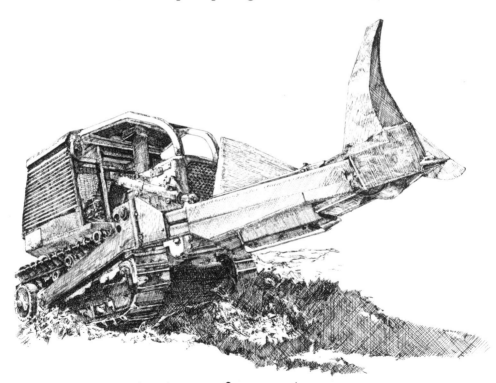

Site preparation by scarification plow

just like a farmer who plows his field. He's not destroying it, he's just loosening up the ground and preparing it for a seedbed, and then he plants it. You aerate the soil to get a better start for the new forest.

"Spraying is another important part of reforestation work. As I said earlier, Douglas-fir likes a lot of sunlight. If for some reason the other vegetation has a jump on the trees — like in our coastal area especially, we have salmonberry and other fast-growing species — those weed species will get ahead of our Douglas-fir. We have to come in and knock that type of vegetation back to give the Douglas-fir a chance. This is what we call our release spray: it releases the growth of the tree, which is able to push through the brush vegetation to reach for the sunlight. A few years later, we will get back the ground vegetation, but we have saved the trees, because by then they are above the brush. We remove an obstacle temporarily.

"After the trees are through the process of competing with the vegetation, when they're maybe twenty or twenty-five feet tall, they start to compete with each other. Of course, we like to start with a relatively large number of trees, because this will give us an opportunity to do further selection. By age twelve or fifteen, the trees have shown their dominance, and we are able to go out there and do what we call a precommercial thinning. We send a crew through and eliminate all those trees that would compete with our so-called crop trees, the dominant trees. While we are there with the power saw, we are able to cut down some of the unwanted species too, some of the weed species that are taking up growing space and using the moisture and nutrients in the ground. By spacing the trees, we open up the stands, and the trees are able to accelerate their natural growth rate.

"In order to expedite the growth rate even more, we are moving in with aerial fertilization. Aerial fertilization is used only in areas where we know we have nutrient deficiencies. For optimum growth you have to have a balanced nutrition program. If just one element is missing, you can have all kinds of other nutrients available, and the trees still won't be able to utilize them. By adding the missing nutrients, we put back a balance, and all the other nutrients become available. Of course, forest fertilization is a new field, and we are still learning a lot in this area. It dates back only ten or twelve years for large-scale application. We normally get pretty good response, but we have found some areas where our response has been poor. We have discovered that certain nutrients are missing, particularly sulphur. That was quite a discovery. By adding sulphur, we are unlocking something and getting a definite response.

"We are examining all timber stands old enough to be fertilized. We are looking at the soil to see what's missing. We can pick out the fertilizers that fit our conditions the best, and then we can order a fertilizer in rail carloads and use helicopters to spread the fertilizer over the various areas. Of course, fertilization costs money like everything else. In forest management you don't want to waste your money. There's no point in fertilizing the soil if you don't have to. A company is out to make a profit, there's no question about that. It's a corporation and it has stockholders. America is a capitalist country, and everybody wants to make a profit. If we don't make a profit, we won't be able to sustain a healthy economy. In our organization, every project has to stand on its own. It has to make a profit and has to be ecologically sound. If it cannot make a profit, or isn't ecologically sound, we have to look at it carefully for possible elimination. All management procedures should be as efficient as possible, and they can only be efficient when we know what we are doing. That's why, for example, we're testing our soils to find out what fertilizers we need. And that's why we run all the tests in our genetics program. We try not to leave any stone unturned."

MOTHERING YOUNG GROWTH: NATURE VS. NURTURE

The tools of industrial forest management can play a vital role in helping young trees to become established. When the seedlings are first transplanted from the nursery, they are easy prey for browsing animals. Deer and small mammals regard foot-high seedlings as tasty, tender delicacies; consequently, such animals like to congregate in logged-over and replanted areas. Since animal damage costs the timber industry in the Pacific Northwest several million dollars a year, the companies have been developing a variety of techniques for protecting the trees.

The ideal solution would be to breed genetic strains of trees that animals simply do not like to eat; but since genetic experimentation is still in its infancy, other methods are being used in the meantime. The trees are sometimes surrounded by individual plastic tubes or other fencing devices. Fencing is costly, but a test conducted by Georgia-Pacific revealed that the growth rates per acre in unprotected reforestation sites were only 40% of the rates found in areas where all the trees were caged. By enclosing the seedlings, mortality rates were cut in half.[9] To the industrial forester, fencing is often worth the extra expense.

An alternative strategy is to coat the young trees with chemicals such as thiram or B.G.R. (Big Game Repellent), which render the tender shoots unsavory to animal palates. Theoretically, the animals are supposed to develop a distaste for the treated trees before they become ill from the toxic effects of the chemicals.

The most direct method of animal control is killing off the foragers. Gophers, mountain beavers, and other small mammals can be trapped or poisoned by setting out toxic baits such as strychnine. Forest managers can decrease deer and elk populations by encouraging hunting. Animals can also be controlled by using herbicides to eliminate their habitats: pocket gopher popu-

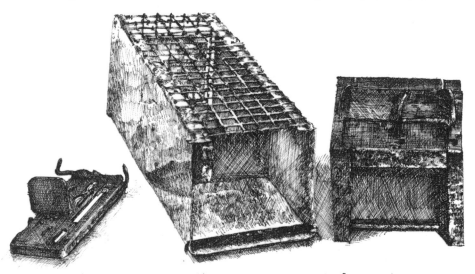

Protection for seedlings: an arsenal of animal traps

lations decline when grasses disappear; mountain beavers tend to leave an area when swordfern, a major winter food, is removed; all sorts of small, foraging animals find it harder to hide from predators when their brush shelter is eradicated.

Animal damage is only one of the many dangers faced by the young seedlings in their struggle for survival. Like any other member of the forest community, the trees from the nursery are subject to disease and insect infestations. They also must beat out their competitors for the available sunlight, moisture, and soil nutrients. Again, the technology of modern forestry can give the chosen seedlings a boost: they can be treated for disease; they can be sprayed with insecticides; they can receive help in their battle for survival by removing the competition. For every crisis planted trees may encounter, there is a chemical substance to aid and comfort them.

The most frequently used chemicals in forest management today are phenoxy herbicides. Developed during World War II and used extensively in the Vietnam War, phenoxy herbicides destroy or retard noncommercial hardwoods, while leaving the more valuable conifers relatively unharmed. These chemicals are applied before planting to eliminate the food and shelter required by undesirable animals; they are applied once, twice, or even three times after planting to eliminate brush competition; and they are used to rehabilitate land that has already been taken

over by brush. Partly because they are used so often, phenoxy herbicides have become one of the most publicized and controversial tools of modern forestry.

ROBERT BARNUM
TIMBER OWNER

"We have a unique situation here in northwestern California. To understand why we need to use phenoxy herbicides, you have to know the historical background.

"When the early logging was done in California, it was done mostly in the redwoods. But when they started logging fir for the great boom in the housing market after 1945, they came into the Douglas-fir stands, and among those stands were great quantities of hardwoods. They logged by tractor because it was more economical, and they had to leave seed trees because of the new forest practices laws. They went in and cut down the best trees, the big tall, beautiful, straight-grained fir with no visible defects. The trees they left for seed trees were the diseased, the conky, the defective. From a genetic point of view, they were doing the worst possible thing: instead of leaving the superior trees as the genetic parents for the subsequent crop, they left the inferior trees. Also, because of the lack of a market, they left the hardwoods.

"Consequently, we are left today with vast tracts of cutover timberland in Humboldt, Del Norte, and Mendocino Counties that were logged that way. There must be several million acres like that all up and down the North Coast and into southern Oregon, and on over into Trinity County and Siskiyou. The hardwood trees that happen to be knocked down sprout like redwoods from the stump, whereas a fir has to come up from seed. So the firs were a long time coming up and were from inferior seed stock. But the tan oak trees and all the various species of hardwoods we have here flourished, and they seeded, and they sprouted, and in the meantime the firs were underneath all this. You have hardwoods that exist because of the alteration of that stand by man, not by nature.

"It's an unnatural situation. The people who say they don't want to use phenoxy herbicides because they don't want to disturb the balance of nature — well, they're just about thirty-five or forty years too late. Now herbicides are used to redress an imbalance.

"They're doing things in forestry now that are really exciting if you're into the business or if you live in a forested country. Up in Oregon and Washington, they're growing timber stands much greater than what grows naturally. It's just like wheat. Back when wheat was discovered four or five thousand years ago, and the guy was picking the ground with his stick putting in little seeds, they didn't grow very much for him. Now, you see pictures in the Midwest where they grow it, and they have these tremendous fields, exports all over the world, and so forth. That's because of the techniques they've developed with hybrids, fertilizers, site preparation, and, of course, phenoxy herbicides. I understand that every acre of wheat in Kansas is treated with phenoxy herbicides.

"The same thing is going on in forestry right now. It's intensive forestry. With the price of stumpage having gone up so much, it's economically possible to do these things, where before it wasn't. Now with these new techniques that have been established, practiced, and experimented with, and are now known to work, they go in and do site preparation before they plant. In some areas where they clear, the grass comes in; they spray for the grasses, which doesn't sterilize the soil, but it knocks the grass back a year or two, giving the little seedling a chance to compete for the moisture in the so-called 'A-Horizon.' And sometimes you might want to leave some brush around to protect from sunlight. Different shrubs take moisture from different levels of the soil. If they take it from deeper levels, you can leave them. Then after several years, you come in and spray. Or you can spray for brush beforehand and then burn, to clear the site, to prepare it, so that these little firs get the sunlight. How much sunlight they get makes a terrific difference in the rate of growth.

"Fertilization is a part of this, too. In the past year, for example, the same companies that are doing the herbicide spraying here fertilized 200,000 acres of Oregon forests. Fertilization of forests has become an economical thing; but you cannot fertilize unless you first control the weed species — otherwise you would fertilize the weed species along with the desired species. That's another reason why phenoxy herbicides are so important.

"It's unfortunate that there's misunderstanding with regard to the health issue. People who have studied it and are very well informed are not concerned about the health issue. They've considered it, and they've determined that it's not a hazard. That's the reason that the Environmental Protection Agency continues to approve its use. There is no evidence to show that it's hazardous to the human animal. If there were, they would immediately take it off. [This interview was conducted a year before the EPA withdrew its approval for 2,4,5-T, one of the most widely used phenoxy herbicides.] There is no evidence of anybody having been injured through the use of phenoxy herbicides, not just this past year but in over twenty-five years of use. It's absolutely safe. I'm concerned about my own health probably more than anybody else's — I wouldn't want to hurt myself or my family — and we have sprayed in our own drainage up in Redwood Creek where we get our water supply. We've sprayed within a quarter-mile of where we take water. We've monitored the water too, and there's no sign of any spray in it.

"People have actually drunk this stuff straight, and it didn't hurt them. I've studied this stuff. I've gone down to the University of California and gone into the laboratories and talked to the people there, and gone up to Portland and talked to authorities there, and talked with local people here at Humboldt State, and all the people who are really well informed say that there's absolutely no hazard or risk with it. You can go out here to the drug store right now, and go in the garden department and buy this stuff right off the shelf in a pint bottle. You can read the label on that: it has 2,4-D and 2,4,5-T. It's exactly the same stuff. It's been out for years. Millions of people each year work with that stuff.

"The irony is that when we go out and put it on the forest with a helicopter, we put it on in the lightest possible dose to do an effective job. There are professionals all the way through: professional applicators; professional pilots; professionals from the Department of Agriculture monitoring it; and our own licensed foresters monitoring the water, watching for the first sign of drift.

You can see the material — if it drifts, you can see it. Then you'd stop if it got too windy. They have very strict requirements. So with all of that, it seems to me that it's obvious and apparent that it couldn't really be dangerous. Doesn't it to you?"

ADJUSTING TO THE NEW FORESTS

The final step in intensive forest management is the harvesting of timber. But this is not really the end, for the cycle will be repeated indefinitely: collect seeds; nurture seedlings; prepare the ground; plant; spray to eliminate weeds, insects, or disease; thin the rows of young crop trees; fertilize the ground; thin again; and, finally, harvest. The process cannot be repeated every year as with food crops, but in most cases it can be repeated two or three times in a century.

A second-growth tree crop at harvest time differs markedly from old-growth timber. There is less rot, less breakage when falling the tree, and fewer unwanted trees to get in the way of the harvesting operations. There is also a lack of the pure, fine-grained wood that takes centuries to develop. The fine grain is a direct result of slow growth — and slow growth can no longer be tolerated by industry. The new trees are not yet old enough to have shed all their bottom limbs, and the wood is therefore knottier. And since each tree is still growing rapidly, a high percentage of it consists of sapwood, which lacks the strength and durability of heartwood. Second-growth redwood timber, for instance yields only 5 to 10% in clear grades of lumber (free of knots and sap), while 60% of the boards from old-growth redwood are clear.[10]

As wood quality changes, markets must be found to adapt to the new products. The California Redwood Association, a marketing organization, once advertised only their clear and construction grades of lumber; today, they emphasize the use of sapwood for paneled interiors and knotty lumber for outside patios, arbors, and fences. The mills, too, must adapt to new-growth trees: they scale down their equipment to deal with the smaller logs, and they process a larger percentage of pressed and glued products to make use of even the lowest grades of wood.

Since the new crop of trees is of uniform age, all the logs are

approximately the same size. This regularity makes it feasible to
develop new machinery specifically geared to the harvesting,
handling, and processing of even-age timber. If the terrain is
gentle enough, the entire harvesting operation can be done
mechanically: a giant pair of shears snips off the trunk at its base;
the tree is immediately rolled through a moveable de-limbing
and bucking machine; large claws called grapples pick up the
logs and load them onto trucks. Hand-operated chain saws and
cable rigging become obsolete, while timber fallers, buckers, and
choker setters are replaced by heavy-equipment operators working at control panels within the cabs of their machines. This sort
of mechanized harvesting show is becoming commonplace on tree
farms throughout the South. The rugged hills of the Northwest,
however, inhibit the movement of such large-scale equipment, so
this area is mechanizing more slowly.

Intensive forest management is generating its own technology.
There are several specialized machines, for instance, designed

Scissoring 16-inch timber: the feller buncher

exclusively for crushing slash and brush during site preparation: the Case Tree-Eater, the Young Tomahawk, and the Kershaw Klear-Way. Hydro-Ax has developed three attachments to fit on a single tractor: a feller-buncher, which cuts and piles the trees; a chain flail de-limber, which removes branches and turns them into mulch; and a brush cutter for site preparation and precommercial thinning. For areas in which tractors and other heavy equipment cannot operate, the slash and brush can be crushed to the ground by a giant steel cylinder filled with concrete and suspended from a high-lead cable.

Explosives can also be used for brush clearing: a canister of propylene oxide is detonated to produce a shock wave that literally strips the leaves and branches from the plants. The frequent use of burning for site preparation has triggered the development of an impressive arsenal of fire ignition devices: shotgun tracer

Tree tongs: a hydraulic grapple loader

shells, napalm grenades, and electrically detonated fire bombs. Aircraft are utilized for a variety of purposes: to ignite and put out fires; to spray herbicides, pesticides, and fertilizers; to survey the land and cruise the timber; and even to harvest the logs. The development of chemicals for intensive management is a whole field in itself. Even computers are becoming widely used in all aspects of the forest products industry: they simulate models for management decisions, they facilitate detailed inventories, they coordinate the logging shows, and they operate the saws at the mills.

This increased reliance on technology has created a new ideology among loggers. Turning trees into lumber is still seen as a challenge by which a logger can assert his manhood, but it is no longer the physical prowess of the lumberjack himself that leads to success on the job. The machine has come to man's aid: it is the extension of his own brute force, the final realization of his control over the forest. Advertisements for the tools of modern logging reveal the psychological equation of heavy equipment with personal strength.

> *International: Logger's Word for Tough*
>
> If you're moving neat little boxes down smooth and easy highways, maybe you can settle for less. But when you're into logging, you've got to move up to International.
>
> Big loads and back roads — or no roads at all — that's where the long-nosed brute of the woods comes into its own. The Transtar Conventional levels hills as easily as a chain saw cuts through kitchen matches; hauls loads that'd shock Paul Bunyan.[11]

Paul Bunyan's Blue Ox has been reincarnated as a modern logging truck with a full payload. His ax has become giant scissors that snip the trees from their stumps. His hands are the grapple hooks that make logs seem like toothpicks. So where is Paul Bunyan himself? He flies high overhead, surveying his plantation with aerial photography. He sits in the cab of his fully mechanized harvester. He programs a computer to manage his scientifically organized tree farm. The task he faces is not how to fall and move a ten-ton tree, but how to grow ten one-ton trees in its place.

Chapter 3

Tree saving: the voice of ecology

THERE is more to the woods, say the ecologists, than resources that are produced, processed, and consumed. The forest exists not just for human use; it has a life of its own as well. The forest is a home for creatures of the wild. It is a symbol of a vitality that goes beyond anything we could create ourselves. Yet now the forest is being tamed: the engineered trees, all planted in rows, are like animals raised in a zoo. As lifetime timber faller Bob Ziak, Jr., puts it:

> I really don't think there are going to be any forests of the future. Forests to me mean splendid old trees, with animals in them, with a variety of trees, snags, and windfalls: everything that it took to create over the hundreds of years that the live trees stood. I feel certain that the forests of the future which the big companies talk about now are going to be nothing more than what they have begun to call them: tree farms. And you won't be able to walk in them because a tree that is thirty-five years of age, he's still bushy, he's still got a lot of little pin limbs. There'll be no majesty; there'll be no cathedral-like feeling.[1]

We ourselves are products of the forest. Important human physical characteristics—binocular vision and prehensile hands—were developed as adaptations to life in the trees. Over time, however, *Homo sapiens* left the woods: less than 1% of the earth's population currently lives under a forest canopy. Still, we maintain a biological relationship with the forest environment. In the words of ecologist Esteban de la Puente,

Healthy forests are an important part of our birthright. Forests fix almost half the total energy of the biosphere. When sunlight falls on the forest, radiant energy is transformed into chemical energy. Photosynthesis turns carbon dioxide from the atmosphere into carbon compounds — the substance of living organisms. In this process of chemical digestion, the forest vegetation releases oxygen back into the air. We, in turn, utilize the oxygen and send back carbon dioxide as a waste product. It's a neat bit of recycling. We coexist. We feed each other with our respective wastes. We're organically linked — the forest and ourselves.[2]

Ecologists perceive a forest not as a farm or a factory, but as a complex biological community with its own economic structure. Ecology — the study of the economics of natural systems — focuses on the interconnectedness of all organisms in a given environment. Over millions of years, a process of mutual adaptation has developed by which each species within the forest community coexists with its neighbors. Today we are trying to reshape this forest ecology, to simplify its structure to serve our own needs. Our real needs, however, cannot be measured in board feet alone. According to the opponents of industrial forestry, our technological manipulation of the ecosystem threatens to deprive the forest of its own economic viability. Unless we change our ways, they claim, we look forward to a spiral of diminishing returns, and ultimately, perhaps, to total deforestation.

GORDON ROBINSON
FORESTER FOR THE SIERRA CLUB

"I believe in multiple-use forestry, which means modification of timber management to provide for the other uses of the forest. Multiple-use forestry consists of managing the forest within the following guidelines or parameters. First, a *sustained yield,* and for that I have my own definition: sustained yield means limiting the removal of timber from a property or administrative unit to that quantity which can be removed annually in perpetuity, where the quantity may increase and the quality may improve, but neither can ever decline. The Forest Service talks about sustained yield, but they're constantly shortening the rotations and decreasing the quantity and quality to sell more timber *now,* to satisfy local, temporary demands by the timber industry.

"The second parameter for multiple-use forestry is to *practice uneven-age management in preference to even-age management, logging only under a selection system.* That means keeping the openings in the forest resulting from logging no larger than necessary to meet the biological requirements for regeneration. There are several species that require open sunlight to reproduce and grow satisfactorily, but that information should not be used to justify cutting forty acres at a crack. You don't have to clear forty acres to let the sun hit the earth. The sun's hitting the ground right here in this yard, and this is only a tenth of an acre.

"The third parameter is to *allow the dominant and codominant trees to mature before cutting them.* Mature in this sense means allowing the trees to achieve their full height. Trees grow like people: they grow up, and then they grow out. Trees have pointed tops, but after a while the upper limbs flatten out, and that is when you call the tree mature.

"The fourth parameter is to *maintain the habitat for all of the species of plants and animals that live in the area.* The reason for this is that there are many subtle interrelationships among the species that live in the forest. We don't know very much about it, but we know enough to know that they are important.

"Take the woodpecker, for example. Woodpeckers are insect-eating birds, as you well know. They're very noisy eaters. Well, woodpeckers need old trees, generally ones with broken tops and decayed hearts, for nesting. It's necessary to have some of these old snags standing as habitat for the woodpeckers if you're going to have them around; and woodpeckers are very important as a control over the bark beetles that kill the trees. I think that the present epidemic of bark beetles in the southern pine region is clearly the result of having eliminated the habitat for the woodpeckers, which reduced the woodpecker population to next to nothing. Not only woodpeckers, but there's a series of species that use these holes. Woodpeckers build them, nest in them, and move on. Then somebody else will move in. These apartments are rented; the tenants change from time to time. But the birds that live in these nests are insect-eating birds, so it's necessary to maintain the habitat for all of the species that occur naturally in the area.

"The fifth parameter for multiple-use forestry is to *take extreme caution to protect the soil*. If we're going to preserve the productivity of the land, we have to protect the soil. When you remove all the vegetation, you expose the whole area to the leaching of nutrients and the erosion of the soil. Soil scientists talk about two basic types of erosion: sheet erosion and gully erosion. There's a tendency to assume that there's no erosion unless you see deep gullies. But sheet erosion is where a whole layer is moved off the surface. The signs are little pebbles spread over the surface: the earth has washed away and left those pebbles. Well, large quantities of earth are lost to sheet erosion following clearcutting. It varies, of course, according to the steepness of the slope and the texture of the soil, but there's some loss anywhere, and frequently there's a lot of loss.

"Another problem is compaction. Going over the land with all this heavy equipment packs it down. Soil is composed, generally speaking, of 50% mineral, 25% water, and 25% gas. If you pack it down and squeeze the gas out of it, you destroy the environment for the creatures that are living off each other in there. In one square foot of earth, there is a population of about 10,000 individuals divided into somewhere between 50 and 200 species. They're all different sizes, and the big ones eat the little ones, just like the fish in the sea. The most densely populated zone in the biosphere is the top foot of earth. Now if you pack that down tight, as you do when you drive over it with tractors or mash it down with these machines which chop up the brush, or you scrape it off with bulldozers and repile it, it's a disaster for billions of individuals.

"Those individuals have all kinds of different functions. There are indications that some of the fungi are able to extract mineral ions out of the large rock, the solid mineral component of the earth. We don't know how phosphorus, for instance, gets out of the rock and into an organic compound, where it becomes available to plants and animals. I think it must be the function of microorganisms. This is a vast area of research that people are just beginning to look at.

"There are some general things that we do know about these interrelationships. The roots of conifers tend to be coarse and blunt compared with those of other plants, and they don't have root hairs. So conifers on their own cannot successfully compete with other seed-bearing plants; but different species of mushroom penetrate the roots of the conifers and draw on the trees for their sustenance, synthesized sugars and starches. Then the mushrooms, in turn, supply the mineral nutrients to the tree. It's a symbiotic relationship called mycorrhizal, and the mushrooms that do this are mycorrhizae. We don't know exactly how these guys work, but we know they're there and we know they're interrelated, and it behooves us to protect them if we're going to survive on this planet.

"Now the mycorrhizae can multiple very rapidly, and I suppose that nature eventually repairs the damage we cause by our logging. But the big question is: can we permit practices that accelerate the rate of destruction? Erosion is a natural process, as foresters will tell you, and they'll use that as an argument to justi-

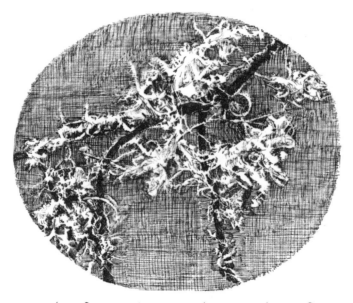

Douglas-fir root hairs with mycorrhizal fungi
(magnified 5x)

fy what they're doing. But it doesn't justify it. There is a natural balance between the rate of formation of soil and the rate of natural erosion and destruction of soil. We've got to maintain that balance, or continually reduce the productive capacity of the earth as we increase our own population.

"And we've got to maintain the balance of the various species. The combinations of species that naturally exist are together because they need each other, because they survived under trying circumstances. I think we can assume that each element plays a role in the common survival of the ecosystems. I don't think we can ignore the hazards of monoculture in the hope of maximizing income and profit. I think that's a delusion.

"The companies are acting irresponsibly by switching to even-age monoculture. But that's nothing new: the timber industry has been irresponsible from the very beginning. They objected to the

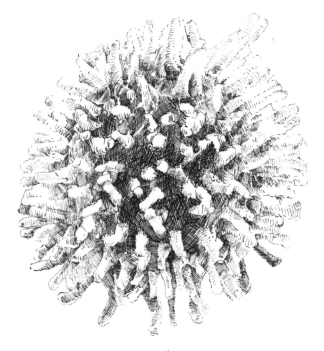

A single spore from the fruiting body of a
mycorrhizal fungus (magnified 3,000 x)

establishment of National Forests in the first place. After the Forests got established, they boycotted them. Then they got to talking about them as 'reserves.' They figured after they ripped off the private land, they could go in and rip off the National Forests. And that's essentially what they're doing. They're all set to liquidate the timber on the theory that they haven't been practicing forestry long enough for their own trees to come to merchantable size. So what we do now is cut off the National Forest land at such a rate that when that's all liquidated, then the young growth will be big enough to cut on the private lands. They're just asking the government to trust them: 'Trust us, brothers, we'll grow timber, only just let us rip off the National Forests so we won't have to cut our little trees until they're big enough.' It's phony, I mean their whole attitude has always been phony. They are planting trees, and some of the big outfits are trying to practice some degree of forestry; but the forestry they are practicing is irresponsible. The timber companies have never been reasonable or fair since the beginning of their industry."

STRENGTH IN DIVERSITY

One of the basic laws of ecology states that: *Other things being equal, the degree of environmental stability is in direct proportion to the number of species living together in the environment.*[3] Insofar as monoculture simplifies the ecosystem by decreasing the variety of species, it would seem to lead to increased ecological vulnerability.

Disease, of course, presents a constant danger to the life of the forest. Tree diseases are many and varied, but most are specific to certain species. When a large number of trees belonging to a given species are concentrated in one area, the spread of any disease which affects that species is facilitated by the dense population of susceptible trees. Epidemics, therefore, are more likely to occur in pure stands than in mixed stands. According to John R. Parmeter, Jr., professor of plant pathology at the University of California:

> One of the cardinal principles of plant pathology is that the greater the purity and density of a plant species, the greater the likelihood of serious plant damage. Forest pathologists long

have called attention to the dangers of monoculture and to the desirability of good species mixtures. . . . The factor of density dependence in disease epidemiology is so generally applicable that it may well be one of the main ecological mechanisms driving plant communities toward the stability of diversity.[4]

In other words, absolutely pure stands of a single species are rarely found in nature for the very reason that they are so susceptible to disease that they have been naturally eliminated.

Insects also tend to focus on specific types of trees. Most insects use a single species or a small group of species as their hosts. Even insects that are general feeders tend to prefer the host species upon which they were reared. In a pure stand of a single tree species, insects which require or prefer that type of tree can find a readily accessible and virtually limitless supply of food. According to a textbook on forest entomology:

It is an accepted biologic law that, other things being favorable, an organism will eventually multiply to the limit of its food supply. As a rule, the more numerous the individuals of a tree species, the more abundant are its insect enemies. When there is an unlimited and convenient supply of a certain species of tree, the stage is set for an outbreak of the insect pests of that tree.[5]

In a diverse environment, on the other hand, there are natural limitations on insect populations. The insects have to search farther afield for their food; whether or not they find it, they expose themselves to natural predators in the process. Tree species that generally occur in thoroughly mixed stands are therefore considered more insect resistant than those that occur in near-pure stands. Jack pine, when it occurred only in scattered blocks on poor soil, was once thought to be an insect-resistant species. Recently, however, jack pines in heavy concentration have replaced the virgin forest species in the Lake states, and these trees are now subjected to repeated attacks by various pests. Jack pine is no longer considered a highly resistant species.

The agricultural analogy that industrial foresters like to draw is helpful in understanding the problems with tree pests in monoculture. Many insects that feed upon domesticated crops — such as the Colorado potato beetle — existed long before the land was tilled. When the vegetation varied, the insects were few and far between. Food in any given area was scarce, so the beetles were forced to travel to new locations where they might or might not find sustenance, and where they would be subject to natural predation. When extensive potato fields were planted, however, the beetles from the surrounding areas suddenly found an abundant source of food that could be had for the taking, and they multiplied wildly, beyond the control of natural predators. To combat the pests, new strategies had to be developed, and so it was that the spraying of pesticides came to be a standard agricultural practice.

As pure stands of trees come to dominate the woods, foresters are starting to face a similar situation. Losses to insects and disease on today's tree farms are not that great — but only because the pests are now controlled with the strategic application of chemicals. The danger is not that infestations and epidemics will wipe out all of our young trees in the near future, but rather that the forests will become dependent on man-made props. Natural balancing mechanisms, such as varying stand composition and encouraging populations of insect-eating birds, are bypassed. The *prevention* of infestations and epidemics is replaced by artificial *controls*. In the words of Frederick E. Smith, professor of Resources and Ecology at Harvard:

> The use of pesticides may seem necessary, even though the manager is aware that it replaces, rather than supplements, the effects of predators and parasites. From then on the system enters upon a long spiral of degradation, as the course of intensive agriculture clearly shows. Agriculture is successful today only because an enormous input of power and materials is used to keep an increasingly unstable system in line. . . . The most perilous aspect of pesticide use is the addiction that follows repeated use. One guaranteed consequence of using pesticides is an increased need to use them again. Not only do pests tend to recover from treatment faster than their enemies, but additional pests are created as other predators and parasites are inadvertently damaged.[6]

TECHNOLOGICAL ADDICTIONS

The dependence on engineered controls constitutes a feedback system: as some controls are used, further controls become necessary. The use of pesticides is only one example; there are several other ways in which our forests are becoming increasingly dependent on human support. When we fertilize the soil, for instance, we supply whatever element has been the limiting factor on growth. The addition of this element, however, enables the trees to utilize more of the other nutrients they need, and the soil is thus depleted of all essential elements at an increased rate. Increased growth is achieved for the present crop of trees, but there will be fewer nutrients available for future crops, and hence even more fertilizers will have to be used.

Another example is the destruction of mycorrhizal fungi during the harvesting of timber. Mycorrhizae aid young conifers by absorbing nutrients from the soil and offering protection from disease, but their numbers are greatly diminished by soil disturbance, burning, and the elimination of host plants. The reproductive bodies of mycorrhizal fungi — called truffles — grow underground and therefore cannot be spread by wind. Fortunately, small mammals such as squirrels, mice, voles, and other creatures of the soil feed upon the truffles and pass the spores through their digestive tracts, spreading them generously from the uncut forest to the harvested areas. Yet small mammals have long been regarded as enemies of reforestation, since they also feed upon the seeds and seedlings of commercial tree crops. Often, they are systematically destroyed, either directly or by eliminating their habitat. With fewer creatures around to spread the truffles, the seedlings suffer from a lack of mycorrhizae. Foresters are just beginning to come to the rescue, injecting the soil in the tree nurseries with the appropriate spores. The ecosystem is being simplified by limiting the numbers of animal pests, but an additional step is being added to engineered reforestation in order to compensate for the function these animals used to serve.

The whole field of forest reproduction has come under man's dominion, and the consequences are potentially momentous. Man, not nature, can now select which seeds will grow into trees and which will not. The seeds that man selects come from a relatively small sampling of parent trees. Variation becomes more limited when a few hundred trees in a seed orchard bear the entire burden of reproduction for many thousands of acres of forested land. Yet it was natural variation that created the giant, straight-grained conifers so highly prized by the timber industry. In nature, only the hardiest seeds are even germinated; in the nursery, almost every seed becomes a tree. In nature, only the strongest and the best-adapted seedlings grow to maturity; under human control, profitability rather than endurance serves as the key to selection — and there is no guarantee that the fast-growing trees will produce a genetic line that is best adapted to environmental stress.

Geneticists, of course, are aware of these dangers and are taking several steps to minimize them. In order to avoid excessive narrowing of the gene pool, they select trees for their seed orchards from a variety of locations and elevations. In order to test for disease resistance, they inoculate different trees with various diseases and select only the most resistant strains. In order to test for environmental adaptability, they keep close records of the progeny that come from each of the genetic strains they develop.

At best, however, all the tricks the geneticists have learned cannot match the ultimate test for environmental endurance: survival for thousands of years in a natural setting. This is nature's test, and the gene pool we have in our untouched forests represents the strains that have passed this test. Nature doesn't select trees according to how easily they can be run through the mill or how quickly they can produce wood fiber. Insofar as geneticists make their selections according to these new criteria, the genetic development of the trees turns away from the simple test of survival. Ultimately, this could mean further dependence on a network of human assistance to provide for the continued health and survival of the desired trees.

Again, the problem is made clear by drawing an analogy with traditional agricultural crops. After thousands of years of breeding for improved strains, most food crops today would be incapable of fending for themselves. In the words of Forest Service geneticist Roy Silen:

> The history of plant improvement is that we usually have been thorough in replacing the original gene pool with an "improved" strain. For example, by the time of Columbus, corn had been so altered by selection done by the Indians that the

original wild corn plant was extinct, and corn's future was entirely dependent upon man: it could not persist in the wild. The history of improvement in wheat similarly has been the refining of the gene pool into one strain after another, none of which could exist without man; each, in turn, was wiped out by a pest. Fortunately, a new improved strain was always in the wings, arising from some resistance gleaned out of the shrinking original gene pool. Today the original gene pool of wheat is reduced to a few small acreages in the "fertile crescent" of the Mediterranean area.[7]

In recent years there has been great excitement over improved strains of rice that might help solve some of the food problems of the underdeveloped countries in the Far East. But the much-heralded "Green Revolution" has run into problems: the rice cannot be grown without the repeated use of fertilizers and pesticides, which the impoverished peasants can hardly afford. Could something similar to this happen to our improved strains of trees? A thousand years from now, will our trees *require* — rather than simply prefer — the application of fertilizers, pesticides, and herbicides? And, if so, will we be in a position to support their addiction? Can we afford to commit our forests to a perpetual dependency on artificial supports?

At this early stage in the development of intensive management practices, we can only speculate about the effects of increased dependency on human engineering, and we can only guess whether foresters of the future will be able to handle the consequences of this dependency. We don't really know where our actions are leading us. This uncertainty makes many ecologists uneasy. Nature, they say, knows best. It is presumptuous, they suggest, to think we can improve on millions of years of natural experimentation. We cannot bypass evolution.

FRED BEHM
HUNTER AND LUMBERJACK

"I belong to this small tree farm organization, and a year ago we made a tour of the Weyerhaeuser tree farm at Cottage Grove. They took us up on the mountain there, and we could look down across it: it was one of the most beautiful sights I've ever seen, thousands of acres of reproduction. But then I looked at it closely and got to thinking about it. It's one of the most depressing things I've ever seen, too. Nothing is going to be there. They'll precommercial thin when the trees are fifteen or twenty years old, then in fifty years they'll clearcut. It'll be like a cornfield. There'll be no snags, no old logs in there. Wildlife — squirrels, birds, and salamanders — will have no place. So many creatures live in the snags and under the old logs. What's it gonna be like without them? Like I say, it's beautiful but depressing.

"My personal opinion is that nature knows what it's doing, and if you interfere too much with nature, you're going to have problems. In nature one thing takes care of the other, and you can't interrupt that without destroying something.

"You can't have it all be just the same. They're breeding these trees for improved yields, but I'm not too sure how that's going to turn out. Of course, they've done that with the grain and the corn and everything else, so I suppose it's possible. But you can overbreed, too. Just like with horses, you can breed up too high where you don't get a good animal. Could be that way with trees, too. But how do you know? You're looking three, four hundred years into the future, maybe a thousand years. I don't think we should put all our eggs in one basket. We should let nature take its course, too.

"I grew up in Wisconsin. I got my first ax when I was seven. I started working in the woods when I was nine. That's when I made my first day's wages. I've been working in the woods about sixty years. I enjoyed every minute of it, too. If I had it to do over again, I wouldn't change a thing.

"When I was in the sixth or seventh grade, the school superintendent came down and asked the kids what their ambitions were when they growed up. They wanted to be presidents, doctors, lawyers, and nurses. Well, we had a neighbor, a Norwegian bachelor who cut cord-wood. He was my idol. I said, 'I'd like to be a cord-wood cutter like the old man.' The superintendent said, 'Son, you must have bigger ambitions than being a cord-wood cutter.' I said, 'No, if I can be a cord-wood cutter like the old man, that's good enough for me.' I've often thought about it; I bet I hit my ambition closer than any of the others did.

"First time I came to Oregon, I saw that pretty water in Blue River and I said, 'Boy, that's the place for me.' So I went and bought a lot, and I'm still here. And I bought timberland. I had 1,700 acres at one time, but when they built the dam they took about 400 acres.

"I've been working for myself since '43, logging my own timber. I run it on a sustained yield — take out so much a year and reforest it. I'm kind of a jack-of-all-trades, so I do a lot of the work myself: fall timber, hang the rigging, set chokers, run loader, run cat. It's my life. I enjoy it. I'm sixty-eight and I'm still doing it. I feel more comfortable working in the woods than pushing pencils.

"I like to hunt. Not so much on my own land, because I like to keep the game around. Of course, the deer are a problem: they browse the trees, the seedlings, and cause a lot of damage. But I don't begrudge it to them. There was deer here before I was. The timber always did survive the deer browsing on it.

"I prefer to hunt up in the high country, up in the Three Sisters Wilderness Area. Pretty country up there. I just hunt for the fun of it. Lot of times, I don't try to shoot anything. I just go around and see what I could get if I wanted to. One year I turned down thirteen bucks. Never did kill one that year. Just looking for a bigger one all the time. Couple of years I carried my bow and arrow with me instead of a rifle. And I do a lot of camera hunting. Several years that's all I did was take moving pictures. I get more of a kick out of that.

"Personally, I think we should keep some land as wilderness. It's important for wild game. I don't think it should all be managed like a tree farm. I'm very disturbed with this idea of trying to cut off all the old growth. I think they should leave a certain amount, especially in areas that are winter range for elk and deer. In a bad winter, that's how elk and deer survive, off that old growth. There's a lichen on the old growth — they just made a study on that down at the university — and it produces nitrogen, about 300 pounds per acre, I think it is. Well, that same lichen is feed for game during a real bad winter. That's nature's way of feeding them: the snow breaks the limbs off that have got that lichen on there, and that's good food for deer and elk. The heavier the snow, the more limbs break off. The brush is covered up on the clearcuts; it's completely snowed over. You get four or five feet of snow and you won't find an elk or a deer track out on a clearcut unit. They'll all be up in the old-growth timber stands.

"Trees don't grow that lichen until they're about ninety to a hundred years old. So if we go with a rapid yield like Weyerhaeuser and these companies are thinking of doing, I don't know. You can raise more board feet by the tree farm method, but it shouldn't all be that way. They should have blocks of old growth and have it perpetual, so that when one crop of old growth gets to a climax stage, you have another block coming to take its place.

"I still have patches of old growth on my land. When they reach a climax stage, they're better off cut, because all they do is deteriorate. I'll let them grow two or three hundred years to reach a climax. I hope I can do it that way. If I get broke I may have to cut them.

"You can't just look at today, you've got to think ahead. That's the reason so much of our private land doesn't have trees on it: people just can't see that they've got to look that far ahead. It's not like growing corn, where you get a crop every year. Weyerhaeuser, they figure fifty years, but that's kind of stretching it a bit. They're figuring out how many cubic feet they can get out of an acre every year. That may be right, but then they're going to have to fertilize, and spray, and all this sort of stuff.

"I don't think these big companies are looking far enough ahead. In the long range, what's going to happen to the soil? In any kind of farming, you've got to have something to build it back up again or the soil goes to pot. Most of these fertilizers they're putting on are just a feed, not really a fertilizer. They're just short-term, a shot to make things grow. They don't really build the soil up. For that you've got to have humus, a certain amount of needles, and stuff like that. You've got to look *way* into the future, not just fifty years."

OVERDRAWN AT THE BANK: SOIL DEPLETION

In the long run, the most important resource of the forest is the soil itself. It is the soil that must provide nourishment for tomorrow's timber. Soil is the fragile skin of the earth, the interface between organic and inorganic elements that constitutes the life-support system of the forest. It takes hundreds of years to create one inch of topsoil by the weathering of parent rock material and the decomposition of humus. In a mature forest, the cycling of soil nutrients is slow and balanced: what is used up by the trees is replaced by natural fertilizers such as rainfall, rock weathering, and the litter that the trees themselves create.

Soil builds slowly, but it can go fast. Nutrients can be removed from the forest in a variety of ways. They can be volatilized, or released into the atmosphere. They can be leached from the top-soil back into the rock mantle. They can be dissolved or suspended in water and carried away by streams. Or the organic biomass of the forest can simply be transported off to other locations.

Intensive timber management practices create nutrient losses in each of these ways. The controlled burning used in site preparation turns nutrients into gaseous forms that are released into the air. Clearcutting exposes the soil and increases the quantity of water that flows through it, thereby increasing the extent of leaching. Road building and other earth-moving activities result in erosion, causing the soil itself to be transported into the rivers and out of the forest. Logging, of course, reduces the forest bio-mass and lessens the amount of organic material that is available for future soil production.

The most natural tool used in timber management is fire. Long before the evolution of human beings, fires caused by lightning were thinning out forests and serving as a check against insects and diseases. Yet even these natural fires left an impact on soil nutrients. Nitrogen, the basis of plant protein, is the nutrient most directly linked to tree growth. Approximately 60% of the nitrogen literally disappears into thin air when a wildfire consumes the organic material on the forest floor. The precise extent of the damage depends on the heat of the fire: sometimes only the litter is burned, but at other times the organic component of the ground itself is destroyed. When piles of logging slash are burned, the heat is intense enough to volatilize over 90% of the nitrogen.[8]

Fire also destroys the microorganisms that contribute to the breakdown of soil nutrients. If hot enough, a fire can alter the wettability of the ground, so the soil will have difficulty in absorbing water in the future. And when the wettability of the ground is lessened, rain will run off the surface of the earth instead of penetrating it. Because of this increased runoff, more soil is likely to be lost to erosion.

Nature, of course, has ways of dealing with these problems. Whether a fire is created by lightning, careless campers, or forest managers is of little significance. As soon as an area is burned, a natural scheme to regenerate the health of the soil is instantly set into action. Fire converts litter and humus into ash, which has a higher pH factor; this decrease in acidity encourages the growth of bacteria that are able to take nitrogen from the air and make it available to plants. Although the nitrogen bank has been depleted, the production of nitrogen is stimulated as the forest begins to build itself back.

The heat of the fire opens many types of seeds that would otherwise have remained dormant, and brush that has adapted itself to a fire ecology quickly comes to dominate the landscape. Many of these pioneer species, such as the several varieties of *Ceanothus,* are also equipped to fix atmospheric nitrogen into the soil. The roots stabilize the exposed ground and help prevent erosion, while root penetration helps break up compacted soils. When the pioneer plants die off years later, they leave the soil enriched and aerated by their elaborate network of root channels. Organic material, created by the photosynthesis of these fast-growing plants, gradually decays and enters the nutrient bank of the earth.

When the pioneer stage of forest succession is bypassed by the application of brush-killing herbicides, the soil is not given the time to go through its normal recovery cycle. By eliminating the "weeds," the soil is deprived of natural fertilizers — which means that more man-made fertilizers will have to be applied in the future. Alder, the ubiquitous weed tree of the Pacific Northwest, fixes between 89 and 286 pounds of atmospheric nitrogen per acre per year.[9] (The variation depends on the age and density of the stand, and the amount of nitrogen already in the soil.) The leaves that alder trees shed each fall are an additional source of nutrients: the litter in a mature alder forest contains about 100 pounds of nitrogen per acre per year, whereas the litter in a coniferous forest contains less than one-third that amount.[10]

What do these numbers mean? Commercial applications of nitrogen range from about 100 to 500 pounds per acre, but these are only applied once or twice in a generation. Alder trees, working year after year, contribute much more than that. One study concluded that alders, through the combined action of nitrogen fixation and litterfall, enriched the soil by a yearly average of 124 pounds of nitrogen per acre over a forty-year period. By the end of that time, the soil under the alder trees had 4,960 pounds more nitrogen than the control area under conifers.[11] Alders might be commercially inferior to conifers, but with respect to the health of the soil they are much more valuable. By suppressing these natural fertilizers, herbicides are depriving the earth of nitrogen, depleting, in effect, the forest soils.

Clearcutting, like controlled burning and the application of herbicides, tends to draw against the nutrient bank of the forest. In a clearcut, all the standing vegetation is cut down. The logs that are removed from the woods can no longer be recycled back into the soil. The debris left behind is generally burned, releasing many nutrients into the air, rather than returning them to the earth. The forest biomass is eliminated in one stroke, and the ground is deprived of organic material to convert into topsoil.

Clearcutting also affects the microclimate of the soil. In a mature forest, water is taken from the ground by the roots of the trees and transpired into the atmosphere through the leaves or needles. When the forest cover is removed, less water is lost through transpiration, which means there is more water left in the ground. But the wetter soil is a mixed blessing. With more water filtering through it, soil losses due to nutrient leaching are increased. Normally these losses would be offset by the nutritive value of the rainfall, but the collection of nutrients from the rain is largely dependent upon root surfaces, and the roots are no longer alive.

Although the ground as a whole is wetter, the surface layer of the soil may actually be drier after a clearcut. Unshaded by trees and unprotected by brush and litter, the exposed ground is subjected to the direct rays of the sun and the blowing of the wind, and surface evaporation is therefore increased. There are no leaves or needles to collect fog from the air during summer drought. The ground becomes hotter during the day and colder at night, wetter during the rains but drier when the sun shines upon it. These greater extremes, in turn, alter the habitat for the microorganisms that normally live in the soil. With fewer microorganisms to help break up the ground, the process of topsoil formation is impeded.

SLIPSLIDING AWAY: SOIL EROSION

With neither trees, brush, nor litter to shelter the ground, surface runoff is greatly increased. The earth is often sealed by a claylike film, which tends to repel water. When rain hits the ground, it runs downhill, rather than down into the soil. In areas of well-established, undisturbed vegetation, surface runoff is rarely more than 3% of the total precipitation. On denuded land, however, over 60% of the rainfall can be lost to runoff.[12] In an experimental watershed in Oregon, surface runoff was increased sixteen inches per year by clearcutting the hillsides.[13]

With more water running over the surface of the earth, more soil is likely to be carried away by the rains. Erosion creates serious problems for the forest environment. In the words of a rancher from logged-over Briceland, California, "Every time it rains around here, a whole lot of real estate changes hands."[14]

Water acts to move dirt in two distinct ways: it detaches and transports small soil particles; and it lubricates the ground to initiate mass earth movements such as landslides. During a heavy storm on bare ground, raindrops can splash more than 100 tons of soil per acre into the air.[15] Some of these detached particles are carried away by the flowing water. As the water rolls off the hillside, it eats away at the earth's surface (sheet erosion) and digs out drainage channels (rill and gully erosion). Erosion due to soil detachment rarely accounts for more than 20% of the sediment that is carried away by streams, but the soil lost is especially valuable. It comes from the uppermost crust of the earth, the rich layer of humus and topsoil that is essential for the growth of any type of vegetation.

The bulk of the sediment in the streams is produced by mass movements. Steep slopes become saturated with water in the wake of logging. The earth has also lost the root systems that held it together. Hillsides at the limit of their "angle of repose" are no longer able to hold themselves up. They literally crumble to the ground in the form of landslides. And landslides tend to perpetuate themselves: the slopes temporarily become even steeper, and all vegetation that might have served as a cohesive force is carried away or buried by the moving earth.

When the landslides reach the streams, the earth material either adds to the flow of water or gets deposited on the streambed, thereby raising the stream channel. In either case, the water level in the stream becomes higher, and the higher water tends to undercut steep slopes downstream from the original landslide. This, of course, creates new landslides, which deposit even more sediment in the streams. These landslides, in turn, raise the water level still higher, which may create additional problems farther downstream.

In this feedback system, the slope instabilities created on hillsides near the headwaters can have far-reaching consequences for downstream neighbors. In Redwood Creek, California, the main channel silted up to a depth of fourteen feet in some places due to logging practices upstream.[16] The channel was forced higher and wider, and it undercut the banks of the newly created Redwood National Park. Trees that had survived flooding for centuries now tumbled into the stream.

The greatest contributing factor to landslides and stream sedimentation is the construction of logging roads. In three small watersheds in western Oregon, sedimentation in the area that had been clearcut *without* roads was 3 times that of the control area; but sedimentation in the area that had been logged *with* roads was 100 times more than the control.[17]

The damage from road building can occur even if the timber is never harvested. When a cut is made in the side of a hill, the normal slopes and weight distributions of the ground are altered —and there is no guarantee that the new slope of the hillside will be stable. The earth that is piled on the outside of the road lacks any vegetative support; the removal of dirt from the inside can undercut the bank and cause it to collapse. Roads disrupt normal drainage patterns, and they can interrupt subsurface water along the inside banks. Water therefore flows where it never did before — and it often travels over ground that has just been rendered unstable. The result, of course, is increased vulnerability to landslides. A study of a logged-over area during a heavy winter revealed that 72% of the mass earth movements occurred in connection with roads, although the road rights-of-way accounted for less than 2% of the land.[18] Another study concluded that a landslide was 300 times more likely to occur along a road right-of-way than in an undisturbed forest.[19]

Landslides, as long as they remain unstable, take up space that otherwise might be devoted to the growing of trees. And so do the roads themselves — particularly if they remain in use between harvesting periods. Trees will grow back on tractor trails that are not being used, but their rate of growth is significantly slower than that of trees growing on undisturbed land. The leader growth on Douglas-fir seedlings in western Oregon showed a 43%

A new road cut: will it hold?

reduction on tractor trails; the growth on 8-year-old Douglas-firs showed a 57% reduction.[20] According to another study, 26-year-old loblolly pines that grew back on old roadways had only one-half the volume of wood of their immediate neighbors.[21]

This effect is partly due to a nutrient scarcity in roadbeds; the topsoil has often been scraped off, and the subsoil that remains has only a fraction of the nitrogen, phosphorus, potassium, calcium, and magnesium found in the adjacent ground.[22] Even more important than nutrient depletion, however, is the alteration that occurs in the *structure* of the soil when it is packed down by tractors weighing anywhere from 10,000 to 40,000 pounds. When the ground is compacted, the soil becomes solid rather than porous: the small air spaces are pressed out, and tree roots find it more difficult to penetrate the hardened earth. Water penetration also becomes more difficult, and this loss in permeability leads once again to an increased flow of water over the earth's surface. A study in southwest Washington showed a 93% loss in permeability along skid trails.[23] In a controlled experiment, four trips over dry soil with a fully equipped logging tractor resulted in an 80% loss in permeability. When the ground was wet, the same damage was done with just a single pass of the tractor.[24] Because compaction creates impermeable soils, the mere existence of heavy equipment in the forest tends to increase water runoff and soil erosion.

THE BALANCE SHEET

The extent of environmental damage, of course, is directly dependent on the specific terrain and the systems used for logging and regeneration. Foresters still find it useful, however, to make some generalized estimates. According to a survey of timberland in various National Forests, approximately 3.4% of the land is permanently removed from production by major haul roads.[25] An analysis of the H.J. Andrews Experimental Forest in the foothills of the Oregon Cascades revealed that 9 to 10% of the area had become unproductive because it was covered by landslides.[26] Although many forested regions are more stable than this, the extension of timber harvesting to steeper, less accessible slopes has certainly increased the losses due to mass movements of the earth.

Nutrient depletion due to clearcutting, burning, and the application of herbicides will also affect future crops. A study in the redwoods of California showed a 41% reduction in nutrient availability five years after clearcutting.[27] A comparison of a mature forest with adjacent cutover areas in Wisconsin revealed that the exposed soils had less than one-third the nitrogen and less than one-half the available phosphorus and potassium.[28] If natural, nutrient-building mechanisms are allowed to take their course, the soil may heal itself in time; but when herbicides eliminate the soil-healing stage of forest succession, the trees are more likely to require the application of man-made fertilizers.

Fertilizers, however, do nothing for the structure of the soil. In tractor logging, about one-quarter to one-third of the land is generally disturbed to some extent by skid trails. When old-growth redwoods are clearcut, ground disturbance covers as much as 80% of the land, because cushioning beds of dirt are prepared for the towering giants when they fall.[29] When mechanical site-preparation equipment is used, the entire surface of the ground is run over by heavy machinery. Whenever and wherever the ground is trampled, some degree of soil compaction is bound to occur. Areas that have been logged by tractor (not including the roads themselves) show a mean loss of 35% in soil permeability.[30] And when the soil is compressed, trees will not grow as well.

These figures do not imply that timber production is about to cease; but they do show that current practices have negative effects on future production. The adverse impacts of industrial timber management are slow but cumulative. The longevity of trees effectively masks many of the dangers inherent in forest practices. Company profits will not disappear overnight, but, in a century or two, the soil may be decidedly less productive than it is today.

IMPACT REPORT: A FISH STORY

Some resources are more immediately affected by current practices. Logging in recent years has had a particularly serious and deleterious effect on the anadromous fish, salmon and steelhead trout, which utilize coastal waterways for reproductive activities. As the fish return from the ocean to lay their eggs, they

often confront impenetrable logjams, which terminate their journey before they can reach the spawning grounds upstream. Even
if their trip is successful, the clear gravel streambeds in which
they themselves were hatched may be covered with sediment from
increased erosion. And, even if they do find clear gravel beds in
which to spawn, their eggs may be suffocated by sediment washed
into the stream during the winter.

Young fish hatched despite these problems may encounter a
drastically altered environment if there are logging operations
nearby. Oxidation of the limbs, twigs, needles, and bark that are
left in the streams can use up most of the dissolved oxygen in the
water. Juvenile salmon, placed in shallow streams saturated with
logging debris from a clearcut, suffocated within forty minutes.[31]
The juveniles also find fewer deep, clear pools to use for cover.
Removal of the streamside vegetation can raise the water temperature more than twenty degrees,[32] and the higher temperatures
tend to decrease the available oxygen still more, while simultaneously increasing the salmon's oxygen requirements. Warm water

Cat in the stream: *but what about the fish?*

increases the salmon's sensitivity to toxic substances; hot water actually kills the fish.[33]

It is little wonder that fish populations have been rapidly declining over the past 40 years, ever since logging and road building reached into the erosion-prone headwaters where salmon and steelhead trout like to spawn. The anadromous fish runs in northern California streams were only one-third as extensive in 1970 as they were in 1940.[34] Along the South Fork of the Eel River, salmon runs averaged over 25,000 adults per year in the 14 years prior to 1952; in the 9 years that followed, the runs averaged only 8,000 per year.[35]

Timber interests respond in several different ways to the charge that logging destroys fisheries. Some plead innocence, arguing that the declining populations are due to commercial fishermen on the open seas. Others admit some wrongdoing in the past, but claim that things are different now. They point to new state laws that restrict logging along the streambeds, and they promise to do their best in the future to keep the streams clear of sediment and debris. Still others, like Dave Burwell of Springfield, Oregon, openly admit that they are willing to sacrifice the fish in order to harvest the trees.

> I wiped out the creek for two miles, but the fish only lived and spawned in the lower quarter mile. So the fish fry when we clearcut. They'll come back in five years. They only live five years in the first place. So we don't destroy. We only interrupt the fish life. We say okay, fish, you can't live here for five years. So we only destroy one crop of fish — they're expendable. To get out that crop of trees, it's justifiable to eliminate one crop of fish. That makes sense, doesn't it?[36]

There is no guarantee a degraded stream will restore itself, no assurance that the fish will return once their spawning habitat has been destroyed. And it is not just the fish that are sacrificed. Several Indian tribes in the Pacific Northwest depend heavily on the fish runs for their sustenance; declining salmon and steelhead populations now threaten these traditional life-styles. Literally millions of sport fishermen now catch fewer fish. And, of course, fishery degradation threatens commercial fishermen with the loss of their jobs.

NAT BINGHAM

COMMERCIAL FISHERMAN AND PAST PRESIDENT OF THE
SALMON TROLLER'S ASSOCIATION

"I think that an intimate connection exists between trees and fish. It's a symbiotic relationship. The salmon need cold, clear water in the streams they spawn in. The trees hold the topsoil up on the hills and keep it from coming down in the streams and choking them up. The trees trap the water in the hillsides above the streams; all through the dry season that hill will still be releasing a little bit of water through springs and feeding the streams cold, clear water. And older trees, for instance, shade the streams and keep the water temperature down. If the water temperature goes too high — like above 75° — there is fatality. Practically speaking, 70° is the upper limit, and the ideal is below 60°. The 50° range is what they really want.

"The trees are transpiring water vapor into the atmosphere, keeping the overall climate cooler and moister, and, in turn, creating a cooler environment for those fish. Salmon are descended from a species of fish that originally evolved as a response to the challenge of the Pleistocene glaciation. All of a sudden you had vast river systems draining the glaciers as they melted, and the fish that formerly lived in the rivers grew to a tremendous size, because there were these huge rivers and giant waterfalls, ice-cold glacial waters all over the continent. As the glaciers retreated, these fish were forced to adapt by going to sea. The salmon is an arctic fish in origin. The water needs to stay as close to that climate as possible for them to make it.

"The salmon are doing a very far-out thing when they come back into these dry hills. The returning fish bring all kinds of essential trace elements back from the sea to the land. They die on the streambanks, and the predators eat them — the raccoons, and so on — and shit them out all over the woods. These trace elements get into the soil where the trees can use them. So there's an interchange going on.

"I have a strong personal commitment to the restoration of natural salmon runs. Biologically and economically, I'm much more comfortable depending on a resource based on the natural world, rather than one based on a whole artificial construct of hatcheries and bureaucracy and everything like that. We're trying to preserve the species, so we can begin to reverse the effects of the heavy environmental degradation that's going on.

"The worst damage to the fishery is done by the incredible amount of sediment that comes from a messy logging operation. The sediment chokes the eggs by being so fine-grained that water cannot percolate through the gravel. In clean gravel the water brings oxygen to the eggs, so when the fry hatch, they'll be able to survive. If the level of dissolved sediment in the stream goes too high, it can be disorienting to small fish. Adult fish can handle sediment-loaded water and navigate through it pretty well, but little fish that haven't been to the ocean sometimes get badly disoriented, displaced, and separated from their food supply during a heavy runoff, when the water is coffee-colored. They just get washed down and starved to death.

"It *is* possible to harvest timber in such a way that you won't ruin the fisheries, and, incidentally, ruin the forests of the future. I feel that forestry and fishing both have a future in the same area, that we can successfully go hand in hand and build a future. But it's going to take some definite changes in policy on the part of the timber operators. As a remedy, what I've been suggesting is to change the Forest Practices Law a little bit. Instead of a $500 fine for a violation, simply deny the operator a permit for another harvest. Make the guy go clean up before he gets another permit. And let that be even-handed: let the private owner feel the weight of that as well as the corporation.

"What's happening now is all backwards. Instead of improving the watershed so the fish will come back, they're trying to create fish that won't even need any wild rivers to spawn in. They call it ocean ranching. The big corporations have gotten the idea of operating hatcheries for profit. It can be done, but it's highly capital intensive. It requires very few workers. And you can put this big food-processing facility right at the hatchery to harvest the fish when they come back in. They use the natural homing instinct of the fish. When the fish smolt, when they change from fresh to salt water, they imprint the taste of the water they're in. So the fish will return from the ocean right where the company lets them loose. But the quality isn't nearly as good as ocean-caught fish. They've burned out a lot of the nutrients in their bodies to get back to spawn. It's like catching a river fish.

"The corporations will try to get the commercial fishermen out of business. They're already doing that. They're lobbying heavily to make it harder and harder for commercial fishermen to fish in the ocean. They say that ocean trolling is most unfortunate because the fish do not grow to maturity before they're harvested. They say you don't have maximum sustained yield, and you can't manage the resource effectively with all those guys out there in their little boats. They want to drive us out of business so we won't be catching their fish. That's the key to it: they think of those fish as *theirs*. We think of the fish as belonging to everybody — and to themselves. It's a public resource, and everybody should have a chance at it. I don't have any problem with a guy going down and taking a fish out of a stream to eat it, but here's a big corporation saying, 'We have the exclusive rights to these fish.' And they fence off the river leading to the hatchery so nobody can catch *their* fish.

"It's like a chicken-feeding operation, except that they don't have to pay for the feed. Our industry wouldn't even have a problem with these guys doing this, if they wanted to buy the feed and keep them there. The big problem is letting the fish loose in the ocean to graze on the forage and consume the protein that's out there for all the other fish. There's a limited amount of feed in the ocean. Their fish will be competing with the fish that are there now. And we will not be able to catch their fish, because they will genetically establish a strain that does not bite a hook. In fact, they will probably try to engineer a strain of fish that does not attain legal size until after the trolling season is over. Their fish are getting smaller and smaller every year already. They're going for this small fish that returns in the second or third year — and it stays at sublegal size until the last two months. It's just like the change in the size of the trees the big companies are growing: they're going for the small but fast-maturing trees that they can harvest more often.

"Up in Oregon, the timber companies are getting into ocean ranching. Weyerhaeuser is heavily involved. They have the biggest facilities going. They love it, because if they can replace the wild salmon stocks with these hatchery-reared fish, then you lose your economic reason for protecting a watershed. That's always been their bottom line: the economy. They claim that timber is more important than wilderness, because it is economically useful. But we're another bottom-line interest group, and environmentalists can always point to how much the fishing contributes to the economy. With ocean ranching you no longer need a fishery in your streams, because all you need is the hatcheries and the ocean. They'd be able to log in whatever way they want. Siltation will not be an issue, because they'll say: 'Well, we're taking care of the consumers. We supply them with lumber, and now we can supply them with fish. You don't need stringent forest practice laws anymore.'"

Fishermen are not the only workers threatened by unemployment. As each phase of timber harvesting has become mechanized, more and more loggers have lost their jobs. The editorials of logging magazines complain that preserved forests and environmental restrictions lead to unemployment, but the articles and advertisements in the same magazines exhort forest managers to save on labor costs by purchasing more machinery — and saving

A choker setter:
will he lose his job?

on labor costs, by definition, means unemployment. Grapple hooks, for instance, make it possible to move logs without hand-set choker cables. When a logging operation switches over to grapple hooks, choker setters no longer have to drag their cables through briars and brambles or cut their hands on the frayed strands — but neither can they collect their paychecks.

Clearly, it is not just the trees and the fish that are affected by modern tree-farming techniques: the lives of loggers and fisher-men are also being changed as technological control over the forest is intensified. Decisions in tree farming have traditionally been made according to assessments of productivity, with little or no attention given sociological impacts. But the impacts on people are real, and they are becoming harder to ignore. Not just the workers themselves, but also the neighbors of the industrial forests are being affected by intensive management. Throughout the 1970s remote regions once thought of as uninhabitable have become inhabited, and the new backcountry settlers have not passively accepted the various activities of timber managers.

HERBICIDES: OF MICE AND MEN

Of particular concern to the neighbors of managed forests is the impact of herbicides on human populations. The use of herbicides has increased dramatically since 1960, and they are now the most widely used chemicals on America's forested lands. Phenoxy herbicides such as 2,4-D, silvex, and 2,4,5-T kill off or retard broadleaf plants by promoting an uncontrolled expansion and division of cells. They also have an effect on animals: several different laboratory tests, conducted both by governmental and independent laboratories, have concluded that herbicides cause embryo and birth defects in chicks, mice, rats, and hamsters.[37] Silvex and 2,4,5-T contain small amounts of dioxin, which an Environmental Protection Agency spokesman describes as "per-haps the most toxic small molecule known to man."[38] Laboratory monkeys exposed to minute quantities of dioxin perished in a matter of weeks.[39]

The manufacturers of phenoxy herbicides claim the substances are relatively harmless. They point to what they call a "perfect safety record" over the past three decades and their own laboratory studies showing little or no damage to animals. How can some scientists claim a chemical is safe, while others say it is not?

The evidence is at variance not because laboratory animals change their behavior for each scientist, but because different questions are asked, and the actual data are interpreted in different ways. Scientists for Dow Chemical, for instance, concluded one of their experiments by finding no significant differences between rats given small doses of 2,4,5-T and those in the control group; other scientists, however, have analyzed the data from the same experiment and concluded that the increased skeletal abnormalities were in fact quite significant.[40] Company scientists have refused to conclude that phenoxy herbicides lead to birth defects — even though they themselves have documented numerous abnormalities — because they define "defects" to include only those problems which directly interfere with the ability of the offspring to survive.[41] Cleft palates or six-fingered hands, by this definition, are not considered defects.

The problem of evaluating evidence becomes even more difficult when dealing with human beings and real-life situations. In 1977 there were hundreds of complaints from northern California residents that they or their animals became ill when nearby areas were sprayed with herbicides. By the time the officials got around to investigating the complaints, however, there was little possibility of coming up with hard, scientific evidence to link the illness to the herbicides. In the words of Ruthanne Cecil, one of the investigators: "Basically all we did was check out anecdotal information five months after people were sick. How can you find out anything five months after somebody's sick?"[42]

Studies on human populations would require large numbers of people to be intentionally exposed to the chemicals, and scientists are naturally unable to resort to this approach. In fact, however, statistically significant populations have already been inadvertently exposed to phenoxy herbicides, and there seems to be some correlation between exposure and various health problems.

Throughout the 1960s large sections of South Vietnam were sprayed with a defoliant called "Agent Orange," a combination of 2,4-D and 2,4,5-T. Populations in the sprayed areas reported statistically significant increases in miscarriages, birth defects, skin rashes, and liver cancer.[43] American GIs exposed to the herbicides have also experienced a virtual epidemic of problems, which parallel those of the Vietnamese residents.

Although statistical data is hard to gather for the dispersed Vietnam veterans, there have been some very precise studies made among chemical workers directly exposed to phenoxy herbicides. In a study conducted on workers engaged in the manufacture of 2,4-D, 44% had liver dysfunctions and 63% complained of chronic headaches while on the job.[44] On six different occasions, explosions occurring during the manufacture of 2,4,5-T have resulted in major outbreaks of chloracne, a skin disease.[45] Agricultural workers engaged in the spraying of 2,4-D (the mostly widely used herbicide on food crops and forests in the United States) were shown to have developed 25 times the normal amount of chromosomal aberrations, indicating that the chemical leads to increased mutations in human beings.[46] A survey in Saskatchewan revealed that 20% of the 3,300 farmers interviewed became ill with headaches, rashes, and nausea after routine applications of herbicides.[47]

The workers in these studies were all exposed to the chemicals at relatively high doses. Does this imply that repeated exposure at low doses is also harmful? Do the herbicides even reach human populations other than the workers themselves?

Studies on laboratory animals have revealed that "the greatest effects of 2,4-D were produced at the low doses administered over long periods."[48] This is precisely the manner in which human beings are most frequently exposed. Once applied, the herbicides make their way slowly into the food chain. One year after a normal application, approximately one-half the dioxin can remain in the soil.[49] Dioxin has been found in the tissues of animals grazing on sprayed land; it was discovered in fish and shellfish three years after the final spraying of Agent Orange in Vietnam.[50] Phenoxy herbicides have been found in the blood, urine, and breast milk of people living close to sprayed areas. Residues have been found in the body tissues of cancer victims who had been exposed to the herbicides.[51]

To the opponents of herbicides, the anecdotal "evidence" drawn from real-life situations is overwhelming — even if the widespread incidents do not constitute valid statistical samples. On a farm adjacent to a sprayed area in Wisconsin, kittens, puppies, chicks, a pig, and a calf were born deformed, the family became ill, and the woman suffered two successive miscarriages. On a sprayed ranch in New Mexico, a farmer lost 38% of his cattle, while 40% of his calves were aborted and many of the live births were deformed. There were eight babies conceived in a small community in Arkansas around the time when 2,4,5-T was sprayed on its source of drinking water. Six of the babies were miscarried, while one of the two live births had a cleft hand and no legs. In a rural community in Oregon, over 40% of the pregnancies resulted in miscarriage when the surrounding area was heavily sprayed; in another community near sprayed areas in California, there were fourteen dead, deformed, or miscarried babies out of twenty-three pregnancies.[52]

According to timber company spokesmen and chemical manufacturers, the isolated incidents are but a series of meaningless coincidences. Dr. Ectyl Blair of Dow Chemical has said: "We think the hazards of 2,4,5-T are really quite small. We've had it in agriculture for thirty years or more and don't believe that the spraying of 2,4,5-T is in any way related to the illness reports."[53] But for those people who have reported ill effects, the circumstantial evidence against phenoxy herbicides seems quite convincing.

ROSE LEE

MOTHER

"Less than a year after I moved to Allegheny [a hamlet in southwestern Oregon], I observed helicopters above me. I became concerned because I knew that my water originated in this one clearcut area where they were spraying. I called a company man to come out and test my water and tell me something about what they were spraying. I was promised the test results and I never received them, but I did receive a nice stack of material telling me how harmless it was. He said that he could appreciate my concern about the sprays, but he assured me that it was no more harmful than ordinary table salt. He said that as long as I

Herbicide spraying by chopper: how lethal is the dose?

lived in Oregon, I'd just have to put up with the sprays. If I didn't like it, the other choice was to move.

"The following summer they had quite an extensive operation. Off and on we'd hear helicopters and we'd see them spraying the area. Now you see, I'm only about a quarter of a mile downstream from this clearcut area on my watershed. I can smell it inside the house when they spray. It hangs in the air for a couple of days. We notice severe crawdad kills in the river; normally the river there is full of crawdads, and following the sprays they'll all turn up dead.

"When my son and I saw them spraying again, I quickly got in the truck and went up there. I was just outraged. They never told me they were going to spray up there again, even though they knew I was concerned. Right above this one large pond there was a whole convoy of trucks and helicopters and a whole crew of men and the chemical tanks. Now there are five houses that take their water from just below that pond. My spring is just a little off to the side from there. Well, I saw them spraying right down in the gorge, right next to the water. I thought, 'My God, there's no way they can keep that stuff out of the water.' I mean they were spraying right next to it.

"Two days later my son and I decided to go take a look up there. It was still, like death. There wasn't a sound. Nothing stirred in that pond. Just two weeks before, we had been up there, and the whole pond was alive with frogs. There's usually all kinds of wildlife up there; but now it was deathly still. There were still droplets of the spray in the air. It was getting on our clothing. You could really smell it. They mix the chemical with diesel oil when they spray it, and rainbows of diesel fuel covered the pond with a thick slime.

"Of course I had not researched this. I was just a layman, just an interested and concerned resident. I didn't know anything about the chemicals, but I wanted to find out more. So I boiled up some mason jars to sterilize them and took some samples to get the water tested. I went around for four months with those mason jars trying to get someone to test the water. I tried every agency from State Forestry on down, but no one would do it. 'Well, it's too expensive. We don't have the facilities to do it. You might try DEQ [the Department of Environmental Quality].' And DEQ would say, 'Well that's not our department. We're not set up for anything like that.' And so it went. Finally I was told by the University of Oregon's Marine Biology Lab that they thought there was a lab somewhere that could do it for $500. Well, I didn't have that kind of money. I think those mason jars are still floating around in the back of a '64 Valiant that's probably been sold and resold I don't know how many times.

"At that point I had to get on with my own harvest. I had a really fine garden. I had moved there so my children, so we all would have quality in terms of the water we drank and the air we breathed and the food that I preferred to raise. I worked the land very hard. It's hard work to raise your own food and do all the canning and freezing.

"That was the summer we had a severe drought. All the reservoirs were down to nothing. And then that fall we had severe flash floods. I mean all of a sudden it just started raining. We had thirteen inches of rain in November. You could see all the sprayed leaves decomposing after sitting there all summer, just washing down into our spring.

"Around Thanksgiving I noticed my daughter was covered with bruises all the time. Of course the kids play outdoors, and they can be pretty rough. I asked Angie, 'How did you do this?' She couldn't remember.

"The holidays came and went and the bruises were there more than ever. So I made an appointment with the doctor, and he checked her platelet count and found it was way down. Platelets are one of the factors that clot the blood, and so her blood wasn't clotting the way it should. The platelet count normally is about 250,000. Spontaneous brain hemorrhaging occurs at about 11,000. Her count at that point was 14,000. So they put her in the hospital immediately.

"She couldn't move. She had blood counts taken around the clock. She couldn't get out of bed or anything, because they were scared she'd bump herself. What was really hard was that she felt fine. You can imagine trying to keep a seven-year-old child perfectly still. If she bumped her head or anything, she could start bleeding spontaneously.

"While she was in the hospital the physician asked me, just as a routine question, whether she had been exposed to any toxins. I don't use really harsh chemicals around the house, so the only thing I could think of was the sprays. I told him about it, and he had her blood tested. And he had my tapwater tested. The herbicides were in her bloodstream, and the same chemicals were in the tapwater. This was eight months after the spray, and the herbicides were still in my tapwater!

"That's when I really blew my stack, because the herbicides didn't belong there. They don't know what the chronic effects of those chemicals will be. They're tampering with life's very essence. I mean we cook with that water, we bathe with the water, I irrigate my crops with the water, we drink the water, everything. There's no excuse for drinking diesel fuel or herbicides or any of the other thousands of chemicals they could pour in our water systems.

"And there's no excuse for those chemicals being in my daughter's bloodstream, because she has no choice in the matter. We don't know how that's going to affect her even apart from the thrombocytopenia, which is the disease of low platelet count. I mean I don't even want her drinking diesel fuel, and I don't want to drink it either.

"Anyway, when Angie was in the hospital her platelet count started picking up; but when I brought her home it dropped again. After that, every time the child came home her count dropped severely. And she could have been dead at any point, or a vegetable at any point, if I hadn't gotten her away from Allegheny.

"At Eastertime I brought her back from the children's orthopedic home just for a quick visit. I was trying to keep my boy in as normal a situation as possible, so he was staying with friends and going to school while I was running back and forth to be with my daughter in the hospital. But I came home and brought her with me, just to take care of the family business and to see my son. Well, they had sprayed again the week before that, and when she came home her count dropped to 10,000. That's below the point for spontaneous hemorrhaging, and it's almost a miracle she's still alive.

"I guess the company won, because I had to move. I could not risk going back anymore. We left the land we loved. We left our community, our friends. And I left my career, too. I was a dance instructor, and I had a good following at the community college. I feel too old to start that all over again in another area and pick up those pieces. At the age of thirty-six with two children I'm driven out of my home, and the kids out of their school. I'm a single parent, and it was a good place to raise kids. Now our life has completely changed. I'm a secretary eight hours a day, typing eight hours a day. So I'm not with the kids as much as I'd like to be. We buy everything at the grocery store. I don't raise our food anymore. This is not what I chose. I can't decide where to go from here. But it's simply impossible to go back home and have Angie's platelet count go down again.

"Angie has had to have surgery. Her spleen had to be removed, a perfectly healthy spleen. And without her spleen, she has no protection at all against pneumonia. Doctors have told me she should be on penicillin the rest of her life. She's had chronic infections ever since her spleen was removed. The spleen is a blood-filtering system, and something — that chemical — was attaching itself to the blood platelets, and the spleen recognized them as foreign and started to destroy them. So the spleen filtered out the platelets. It's like leukemia, in a sense, or like cancer.

"Which reminds me: in the other five houses that take their water from that vicinity — where the spray had gotten into that pond — in each of the five houses there's cancer. Two people in one home, the man and his wife. In the last four years, three people are dead. Colon cancer, uterine cancer. The cancer cases are from people of all ages — some young, some older. And the people there eat very well. They live healthy lives. This kind of cancer epidemic didn't exist before the spraying.

"When I first found out that the herbicide was in Angie's blood, I put notices all around town for a public meeting at the schoolhouse. I talked with some neighbors, and we invited people from the company and all the agencies that have anything at all to do with this stuff. And most of them came or sent a representative. All the media came. Well, it was an incredible town meeting — the interest and the cross section of people who came. There were folks who went back in that area three generations. There were the cattlemen, the hunters. There were the loggers, the senior members of the loggers' union. There were the longshoremen. There were all the people who made up the community. People filled the place, and they stood four feet deep outside, with the windows open so they could hear.

"This was in February, and it got kind of cold outside. But it got hotter and hotter for some of those guys inside. Like one hunter said: 'I do a lot of hunting in this area. What happens to the animals that graze on this stuff? Does it get into them? And what about the meat we eat and the fish we take from the river?' So the head of Fish and Game stood up and said, 'Oh, no, it's absolutely harmless. It won't hurt you at all. It's nontoxic to fish and wildlife.'

"Well, Henry Crump is an old logger, probably as old as some of the virgin timber in that area. He lives off the elk and the bear and the fish. He had shaved for the meeting and he sat there with a gleam in his eye. And when the Fish and Game man said that about being nontoxic, Henry picked up an empty herbicide barrel and walked up in front of all the cameras and set it down on the table and said with a voice about as loud as a chain saw, 'Now I found this old can in the river, and this says right here: *Toxic to fish and wildlife. Not to be used near domestic water supplies or irrigation channels.*' Of course that was a direct contradiction. The main loophole for this whole thing is they say it's 'safe if used according to manufacturer's label.' But the manufacturer's label says don't even put it in an irrigation channel, let alone in people's drinking water.

"At that meeting we really opened a can of worms. Our State Forester got up and said, 'We monitor every single operation that we do.' And one of his main staff people got up and said, 'Now you know, Clifford, that those bottles are still sitting in the basement from last year. They have never been analyzed.' The samples were taken, but only three of fifty were actually sent in for testing. They don't have the facilities, they don't have the funds, to test.

"One of the places that they did test was the Girl Scout camp. And it came back from the state that the water was contaminated. When the caretaker went up to the spring two days later, the salmon were all dead, belly-up. There's a nice, thick stack of affidavits about all this.

"After the company official got up and told how harmless it was, a logger who was a foreman stood up and said, 'That's a goddamn lie. I was on a unit two years ago. The boys that were working called up to the company and said, "We've been sprayed down here. What do we do?" The guy told them, "Go back to work. Stuff won't hurt you." They went back to work. They all puked out their guts. They were home sick for two days. They came back three days later and there were three dead cow elk in the area. Don't tell me that stuff doesn't do any bad things.'

"We had three rare spotted owls, which are on the endangered species list, turn up dead. Citizens from the area would bring in the birds to State Fish and Game. The first one was brought in after they sprayed, and we asked them to test it. There wasn't a mark on it. The following year, after another spraying incident, we brought in another bird and asked them to test it, and we said, 'Whatever happened to the bird we brought in last year?' 'Oh, that bird is sitting on so-and-so's desk — stuffed.' They never tested it. They never took it seriously. They laughed when we told them about the crawdads dying. But they quit laughing once they found that the herbicide was found in Angie's bloodstream. They know that that particular disease has been associated with industrial toxins before, to lots of things with benzene rings. And this chemical also has benzene rings. And maybe they know that Angie's disease has been linked to herbicides in tests on monkeys.

"But it's like saying that cigarettes cause lung cancer. The companies will fight it all the way; they'll pull everything they can. It's very difficult to prove your conclusion, even though it seems so obvious. But what do you do? Do you let this stuff ride until there's just too many tragedies?"

Chapter 4

Tree growing: the voice of holistic forestry

NATURE, if left alone, will certainly grow trees. The impressive coniferous forests that European-Americans discovered in the Pacific Northwest less than two centuries ago bear witness to nature's bounty. But the ways of nature are slow; to some even tedious. Since we want more timber from fewer forests, we grow anxious to get on with the business of growing trees. We substitute genetic engineering for evolutionary processes; we bypass natural stages of forest succession; we even cut the timber before it has reached its optimum productivity. By intensively managing the forests, we hope to straighten out the kinks in the meandering ribbon of time. The problem, however, is that we don't always know what we are doing. In the name of making improvements, we often create new obstacles.

So what should we do? Should we sit back and wait for nature alone to heal the scars that we ourselves have created? Or should we grab destiny by the wings and hitch a ride through time by altering the genetic selection of trees, eliminating unwanted brush with herbicides, and cutting the adolescent timber as soon as we can find a market for it? Fortunately, these are not the only alternatives. There are ways of tinkering with nature that complement, rather than contradict, natural processes.

Holistic forestry, as it is sometimes called by its proponents, aspires to work *with* natural laws rather than *against* them. Reacting to the excesses of industrial manipulation, self-proclaimed holistic foresters attempt to manage land according to the basic ecological principles of balance and diversity. Instead of simplifying the ecosystem by eliminating unwanted elements, they prefer to utilize all elements to the best advantage of the entire forest. Like industrial foresters, they are interested in the production of timber; unlike industrial foresters, they treat the forest as a complex, interdependent system with a life of its own.

The prime concern in holistic forestry is growing trees. But holistic foresters are not interested in just planting trees and spraying them with various chemicals. Instead, holistic foresters offer a sweeping program for environmental repair. Since much of our forested land has already been severely damaged, holistic foresters feel we must take steps to heal the ground that supports the trees. We cannot expect the land to produce to its full potential, for instance, unless we can halt the erosion that washes away the topsoil. We must also replenish depleted soil nutrients, but this doesn't require chemical fertilizers. "Green-manure" trees, such as alder or *Ceanothus,* fertilize the ground as they grow.

Since holistic foresters believe in balance and diversity, they try to maintain a variety of species and ages in the forest at all times. They are reluctant to remove the entire forest canopy at once, since this will alter the microclimate and drastically change the ecology of a site. When they do harvest the trees, they devote their attention to a careful and complete utilization of all the material they remove; they try to find some use for the slash, the low-quality logs, and the species of lesser commercial value.

Holistic forestry treats each specific site according to its own needs. The ecosystem is maintained in its basic form, but is modified here and there to bring it into harmony with human needs. In some spots, for instance, a brush cover might be maintained to stabilize the soil or to replenish valuable nutrients, but just a few feet away, where the soil is in better shape, the brush might be cut back to allow room for conifer seedlings. The premise is that each place in the forest is absolutely unique, even though there are basic ecological rules which must be adhered to if the forest as a whole is to prosper.

The concerns of holistic forestry extend from the microcosm to the macrocosm. On the one hand, each individual site has its own specific requirements; on the other hand, entire forests are dependent on the fate of areas that are geographically distant. It is known, for example, that the water table temporarily rises and runoff increases in an area that has been clearcut. When there are no trees left standing, there can be no evaporation or transpiration from the leaves or needles, and the ground itself must therefore receive the entire rainfall. The hydrological effect of clearcutting, however, is not limited to the specific site. Since evaporation and transpiration are diminished, very little moisture is released back into the atmosphere. When a storm passes through, there is no "recycling" of water directly back into the rain clouds. Consequently, the storm will play itself out sooner when there is little or no vegetation covering the ground. A clearcut area receives more water, but that means that there is less water available to other sites in the leeward path of the storm. In this manner, extensive clearcutting in the coastal ranges of the Pacific Northwest can actually result in a lighter rainfall for the Cascades or the Sierra Nevada. Ultimately, deforestation around the world could alter the climate of the entire planet.

It's difficult to translate this sort of macrocosmic realization into practical terms. It is unlikely that a manager of a small tract in the Coast Range, for instance, will alter his plans because of the very slight effect his actions might have on some trees in the Cascades or Sierras. But holistic foresters are concerned with whole systems, and the ultimate system is the earth itself.

To be consistent with its own tenets, holistic forestry can only be perfectly viable if practiced on a global scale. In fact, however, holistic forestry is rarely practiced on a parcel as large as forty acres. The problem, of course, is that many of the earth-healing practices advocated by holistic foresters would consume large quantities of capital. To the great extent that the earth has already been damaged, proportionally great amounts of time, energy, and money will be required to remedy the situation.

GERALD MYERS
HOLISTIC FORESTER

"My dad, granddad, and great-granddad were Douglas-fir gyppo loggers. They homesteaded Oregon about 1913. When great-grandpa and great-grandma came, they had a huge family, all boys; so they homesteaded the whole valley up there. They logged it in the 'teens. That was all horse logging. They didn't do badly with it. In 1939, before World War II, the land was really in good shape, because they left the trees on the ridges for seed trees, and they didn't muck around the streams. A horse just doesn't muck things up the way a cat does. After the war, the two boys that were left got into contract logging, working out all around the Northwest.

"Growing up, I went to eight different grade schools. We took off after the war with a gyppo logging outfit, up the Columbia River gorge and up in central Washington. When I was in high school, I started working with them in the summer.

"There's a funny cycle about working in the woods, and I'm still affected by it. You bust your ass in the spring, as soon as you can get in, as soon as the roads are hard enough to drive on. You work six days a week. You work dawn to dark, literally. I remember getting up at 3:30 in the morning in the middle of summer, when it was still dark, ready to crank the chain saw up at first light.

"We went broke. In '48 a lot of people went broke up in Washington. It was a real bad winter, started real early. Unemployment ran out about December. It wasn't welfare, like it is now. It would run out around Christmastime, and things would get pretty lean along about January and February. Which is really screwy, when you think about it. Those guys should have been planting in the winter. They might not have made as much money, but they would've kept in shape. Every spring all the loggers would go through this big number for the first few weeks. Aches and pains all over. Bad news.

"So I grew up in that — growing up in the woods and seeing them disappear. None of the loggers wanted their kids to work in the woods. I mean they all liked it, and it was all good, honest work and everything. But none of them wanted their kids to do it. It was ass-busting work with a cycle of boom and bust.

"My parents didn't want me to work in the woods either, so they trotted me off to college. I studied business at Oregon State; I was going to be a businessman in the city. That's what they wanted. And I did it: grad school at Cal, Berkeley; the army; then a systems analyst. Thirteen years in the city. Then I came home, back to the woods, in 1970.

"When I came out here, I came with a different attitude from what a lot of people have when they come from the city. First of all, I could see this land. I knew what it really *looked* like before the logging. And that's a peculiar kind of curse. I couldn't totally enjoy it in its present condition, but I could see its potential, what it *could* look like. A lot of potentially productive forest land wasn't producing anything except sediment in the creek.

"People come out from the city and think it's all very beautiful, but they don't really know how to *look* at a hill. They look at the trees and say, 'Gee, that's pretty,' but they may be looking at an area that's just been ravaged. Too many hardwoods were released when they took the old-growth fir. The cats screwed up the microhydrology. But there's still a lot of potential. The site may have been set back one productivity class, say from a III to a IV, but it can be brought back.

"The problem with bringing it back is that there's no money in it for us. It's all for the next generation. Somehow we must find ways to alter that. People can't *see* what it was, what it is, and what the potential is. So a lot of it is just education. What I've come around to is just doing a lot of the work so people can see it, before and after.

"The other difference between me and the people who came from the city is that I don't hate the loggers. It's real easy for the kids to come out here and say, 'Those stupid-ass loggers, those dumb rednecks.' That's bullshit. Those people were just trying to make a living. For a lot of them, it was the only way they could live out in the country. It was the only work available, and it's good, honest, hard work. And what were they doing? They were filling a demand. It wasn't the loggers' fault they were ripping off the woods. The cities wanted the wood. L.A. was built with the wood out of this watershed here, and from Salmon Creek, and all around here.

"It's almost as if our civilization can't afford forests. We have to cut them down, get rid of them. It's not a new story. Forests are the fastest-disappearing ecosystem on this planet. There's a lot of wealth there, and it's real easy to exploit it. But once you exploit it, if you don't take some of the wealth and plow it back in, as with farming or any continuous operation, it's going to decline in productivity. It's going to go back in ecosystem succession from coniferous forest to hardwoods to brush to brush-burn rocks to sandstone to desert.

"It's happening in Pakistan. It's already happened in Lebanon; the cedars of Lebanon of Biblical times are gone. Northern Africa used to be forest. They took the forest off in the early Egyptian dynasty. It was probably all gone by about 1,000 B.C. Then they grazed the hell out of it with goats, and now it's desert. Yugoslavia was all forested, and it was deforested in a very short time about 280 B.C. to build the Roman fleet. All the soil that built up over eons washed down into the valleys, and Yugoslavia is still deforested. Rome and the hills around Italy: the same thing. All around the Mediterranean: the same thing. The cycle is clear.

"People who are not able to learn from history are condemned to repeat it. We're seeing that right here in California. You start with a healthy forest on this sedimentary sandstone. You cut down the trees and run cats all over the place and screw up all the old drainage patterns. And you don't replant. Then the heavy rains come in the winter, sixty or eighty inches of rain, and take the topsoil that's taken thousands of years to build up, and it runs down into the creek and screws up the fishing. The brush comes back — a lot of it. It becomes increasingly combustible. So some hot summer a fire comes along and burns it off. Then brush comes in to replace the hardwoods, a chaparral species like *Ceanothus* or manzanita. Then you get into a brush-burn cycle, and it's real hard to get a forest back in there. You get enough of that and change enough microclimates, then your macroclimate starts to change. It starts to dry up. The rain just runs off the soil, and you get this progressive drying out.

"About three or four years ago, I started daydreaming about taking this damaged environment and trying to put it back together. That was the first time I started thinking holistically about this place, about this nineteen-square-mile watershed I live in. I thought about putting together a project that would take the out-of-work rural poor (which we have a lot of) and the damaged environment and put them together. A labor-intensive environmental repair.

"I started a volunteer work group. The idea was to heal and maintain the long-term productivity of Redwood Creek watershed lands. Actually, it isn't just a project — it's a direction to take in life. I'll be at it the rest of my life.

"We got a little C.E.T.A. [Comprehensive Employment and Training Act] grant to put together an environmental atlas with the basic physical information about the watershed: population, land ownership patterns, rainfall, sediment yields, and so on. Then we got another C.E.T.A. grant for holistic forestry training. The vehicle we used for the training was to construct a fuel break. We taught a lot of specific skills but also tried to teach appreciation of the forest as a *whole,* not just as a monocultural crop.

"To heal a watershed, everything is site-specific. You're looking at land that has had a varied history. There's no general rule-of-thumb. You have to look at *this* piece of land: the soil, exposure, and previous land-use patterns. How badly has it been eroded? What kind of tree cover does it have?

"There are specific skills involved. One of the interesting things about holistic forestry is that when you look at the whole site, you find that a lot of the work fits together. For example, in timber stand improvement you generate a lot of waste material, and that can be used for erosion control, in contour wattles and check dams. You cure two problems at once. But most government programs and industrial forestry contracts, too, talk about a contract for just one thing — a contract for timber stand improvement, or for conifer release, or for this or for that. And you get into some of the craziest goddamn situations that way. For example, suppose you are doing a timber stand improvement project; a gully runs through the forty-acre tract you are working on. You may be required by contract to burn the stuff you cut, instead of using it to stop erosion in the gully.

"Holistic forestry *can* be done — and it *has* been done right here in southern Humboldt. From any high point in southern Humboldt, you look out at the Tostens' ranch. Their land sticks out like a green thumb. There were two differences. One, they owned

the land and intended to stay there. Forestry was part of what
they were doing, along with sheep ranching and all the rest. They
looked at the land first and figured out what it could provide.
The other thing they did was replant. It's so insanely simple, it
almost drives you nuts. Now, their land is real healthy, real productive. You can see it, right down the borderline where the
gyppos logged up one side and then split. But the Tostens intended to stay, and they did.

"The concept of holistic watershed repair is extremely logical.
There's a natural, ecological, social, geographical unit. It's a
whole piece, this watershed. When you work on it, it has favorable effects all the way down. The problem with it, of course, is
there ain't no money for it. There's still a lot of money to be made
exploiting land, and there's not a dime to be made in repairing it.
It's enormously shortsighted. Can't we afford healthy land?

"Ideally, what I'd like to get is $1½ million for a ten-person
crew, full-time. We could fix all nineteen square miles of Redwood Creek that way. I'd use more controlled burning and do a
lot more planting. Live plantings, stakes, tan oak acorns, anything that will work on a particular slope. Again, it's site-specific.
You look at the slope. Are you really down in the subsoil? Is that
all you've got? Is it wet in the winter? How wet? Does it dry out? Is
it southern exposure or northern? All these things will vary. You
figure out what will grow there based on local conditions. If all
else fails, you plant tan oak acorns. Anything to get cover on the
ground. Once you get that initial cover, then you can come back
with a higher-level or higher-succession plant. If forest was there
before, forest can be there again.

"If you ever expect to get any rural public works financing to
concentrate on the long-term productivity of the land, you're
going to need decent cost-benefit ratios. You have to be able to
do the work on the land for no more than the market value of the
land per acre. Rather than pile and burn the waste material from
brush clearing, for example, why not run it through a shredder?
Why not sell the chips?

"We need more infrastructure. We need a small electric plant
that runs off steam, and then we'll have a market for chips. You
give me a market for chips, and I'll give you a market for all the

excess hardwoods. You give me a market for all the excess hardwoods, and all of a sudden they're not a problem anymore,
they're a resource. They could be utilized fairly close to home, so
you wouldn't have to go through a big fuel-transportation hassle.
You'd wind up exporting electricity. They'll spend $2½ billion
on a nuclear plant, and they won't spend the hundred million for
a turnkey steam generation plant that you could set right down in
the Garberville area. With the present-day steam-generation
capability, it doesn't have to be polluting at all. They inject oxygen, and they keep the smoke real clean. Once you get it up to
temperature, it doesn't smoke. It's real simple; there's nothing
esoteric about it at all. But it's just too small. It just isn't sexy
enough or something."

FORESTRY BY HAND: MANUAL RELEASE AND SITE-SPECIFICS

The challenge of holistic forestry is to create alternatives to industrial monoculture that are economically viable. Unable to depend exclusively on government grants for environmental repair,
holistic foresters must seek out techniques that can simultaneously heal the land and turn a profit. Is that an impossible task?

In fact, nature has its own healing mechanisms that might well
be modified to suit human needs. Consider the properties of the
red alder tree, for instance. Alders thrive on disturbed soil, and
there is plenty of disturbed soil in the wake of most logging operations. As a pioneer species, alder is a quick-rooting tree that can
penetrate compacted soil and loosen up the earth for successor
species such as Douglas-fir. The alder roots serve to stabilize
slumping soils. The leaves transpire excess moisture from the
ground, and this, too, helps to prevent mass earth movements.
The roots also host nitrogen-fixing nodules, and the leaves that
the alders shed each fall turn into one of the richest of all tree
litters in the United States. As an added benefit, alders host certain mycorrhizae that minimize the incidence of root diseases, not
only for themselves, but also for neighboring trees of different
species. The red alder, in short, is an ideal healer of damaged
land; it simultaneously penetrates, stabilizes, and enriches the
soil, while providing a kind of inoculation against disease for
other trees nearby.

Why, then, do industrial foresters treat the alder like a weed and try to eradicate it from their forested gardens? Because alders grow quickly, too quickly for the comfort of commercial tree farmers. Alders tend to dominate many sites, suppressing more valuable Douglas-firs. The industrial reponse is to eliminate the alders and "release" the conifers. The soil-doctoring trees are removed without being allowed to practice their medicine.

One alternative would be to let the alders "do their thing" and heal the soil all by themselves. I call proponents of this point of view "naturalists." According to the naturalists, the conifers will get their turn in due time. Alders grow quickly, but they die quickly, too. After fifty or sixty years, the alders start to fall down and open the canopy for other types of trees. Gradually, more valuable conifers will come to replace the alders. The naturalists point to the fact that this system worked well for thousands of years, but they ignore the economic realities of today. Allowing prime commercial forest land to remain without sawtimber for more than half a century is not economically viable.

Red alder: weed
tree or natural fertilizer?

As long as we present the issue as alders versus conifers, the alders will continue to be mercilessly eradicated. Fortunately, however, there are other management alternatives that combine the interests of the naturalists and the industrial foresters. One system would allow alders to prosper for about ten or fifteen years in the wake of logging operations. During this time, the trees could be expected to grow to a diameter of four to five inches and a height of thirty-five to fifty feet, accumulating a significant biomass that could then be harvested for use as energy. Interestingly enough, the alders' nitrogen-fixing capacity is greatest in the early years of growth and on severely depleted soils; after they have enriched the soil for ten or fifteen years, their nitrogen productivity tapers off.[1] By allowing the alders to thrive on their own terms for a relatively short period of time, foresters would reap most of the benefits from their soil-healing properties. And they would reap economic benefits as well. If we assume a significant increase in the demand for wood as fuel in the near future, the young alders could pay their way out of the woods, and the land would never really be out of production. As petroleum becomes harder to come by and chemical fertilizers become more expensive to manufacture and apply, foresters will save additional money by using natural rather than artificial methods of fertilization. Essentially, tree farmers would be planting a "green manure" crop, just as traditional farmers have done for years. There is nothing really new or radical about the idea of rotating crops to save the soil from eventual exhaustion. Indeed, it is strange that the timber industry, committed as it is to the agricultural model, has not yet embraced the idea of crop rotation.

Another alternative suggested by a holistic approach would allow commingling between alder and Douglas-fir. Since both species tend to do better in proximity to each other,[2] we could have our cake and eat it too. The alders would provide both nourishment and protection for the young Douglas-firs, while the mere existence of the Douglas-firs would provide an economic raison d'etre for the alders.

Unfortunately, this is a difficult method to implement. Since the alders grow far more quickly than the Douglas-firs during the first few years, they would have to be held in check, yet not totally eradicated. There is no way for area-wide treatments, such as the aerial application of herbicides, to maintain a continual balance between the species. Instead, the balance would have to be maintained manually. Laborers would have to move through the entire forest, cutting down a few trees here and a little brush there in order to obtain the desired result: a genuinely mixed stand. Choices would have to be made on the spot by individual workers. This is labor-intensive, site-specific forestry; according to some holistic foresters, it is the only real alternative to the mechanized, computerized techniques that now prevail in the timber industry.

RICK KOVENS

MEMBER OF GREAT NOTIONS, A REFORESTATION COOPERATIVE

"A couple of years ago, some people in Oakridge wanted to have a 'brush-in.' The government planned to spray herbicides on some acreage that was right on these people's water source in Salmon Creek. There was a lot of controversy about it. So a bunch of people went up there and cleared the brush by hand right before it was to be sprayed by helicopters. None of these people were cutters. They went up there with machetes and kids and everything. It took them two days to do twenty acres. Apparently they did a terrible job, but it was politically effective.

"So the next week, some environmentalists down in the Cottage Grove district wanted to do the same thing. We called them up and told them we'd go down there as a crew and cut the unit with saws. It would still be a 'brush-in,' but this time we wanted to do it as a professional crew and not just as citizens. It worked. We went through thirty-five acres in about two hours. It immediately dawned on us that there was a lot of money to be made in brush clearing. So far this had been an environmental issue and a political issue. I realized at that point that it was actually a forest labor issue, an economic issue. If people who were serious contractors would get involved in this, it would make it more viable silviculturally and it would also be quite lucrative.

"That summer the Forest Service started letting out contracts. We got our first one for 130 acres, then we got another one at the end of the summer, one in the Coast Range and one in the Cascades. It worked out really well. The Forest Service and BLM let out 3,500 acres all together. They had competitive bidding on the open market, and, as far as I could tell, it was a really positive thing.

"I started to get more interested and looked at what foresters and the industry people were saying about manual release. The more I read of what little there is, the more I realized that what they were saying was absolutely bullshit. I kept thinking, 'Why are they so irrationally against this?' Then I realized that there was no good research ever done on it. The research was all anecdotal. There have only been two reports on manual release. One of them I kept seeing referred to everywhere. It was quoted in all the Environmental Impact Statements. So I finally got hold of a copy of the report. It was three-quarters of a page long, and it said something like this: 'In 1977, we asked a contractor to do fifteen or twenty acres of brushing, and he wanted $500 an acre. We said, "Forget it." So we got a Youth Conservation Corps crew out there — kids — and it took them a thousand hours, and it cost us five hundred bucks an acre. They killed all the trees. And me and Joe went out there the next year, and by God the brush had grown tremendously. It was three feet tall.' Then the last paragraph: 'The conclusion is that brushing is ineffective and uneconomical.' And that line was quoted everywhere.

"The other report, the biggest report on manual release, goes like this. This guy Brown has a county forestry program, and they do all their work with C.E.T.A. employees and county employees. They have real scrubby, unmanaged land, stuff that was cut in the forties. So this guy takes $25,000 and sets up his study. Then he finishes his report and goes to Washington, D.C. and every forestry convention that he could go to with a slide show and everything. This is *the* report on manual release. This is the accepted authority. This is what everybody uses as their data base.

"The first impact he comes up with in the report is physical damage to released conifers. He says that only 70% of the trees aren't smashed or nicked or cut. Well, if I went out as a contractor and did that, I wouldn't get any money. I've got to get 90% or better perfectly done work or I don't get paid. And no brushing contractor ever failed to get 90% out of all the 3,500 acres that the government let out. Those were 21 different contractors. His figures are completely irrelevant, because he's not using professional contractors, he's using a C.E.T.A. crew.

"He's got 30 acres total in his report, and here's his cost per acre (this is what they use when they write in the E.I.S. [Environmental Impact Statement] reports): 'Manual release costs $550 an acre to $1,260 an acre — reference, Brown.' Thirty acres! Well, we studied the contracts from all 3,500 acres that were brushed last summer in Oregon, Washington, Idaho, and California, and we came up with an average cost of $106.85 per acre. The range was $32 to $366. And he's going around the whole United States telling everybody it costs $500 to $1,200.

"I thought that somebody had to do better research than that. Somebody had to get the information together to prove that brushing is an effective alternative to herbicides. So that's how our research got started.

"What we're doing now is trying to develop some idea of the most appropriate uses for brushing. We want to find out where it will really do the most good. For manual release to work, you should have a unit where the crop trees are ready to release, about four feet high. If the seedlings are too small, they practically get smothered by the piles of brush. Or sometimes they stick you on a unit where the brush is twenty-five feet tall. There's just too much volume to deal with; they might as well burn it and start all over again. They give you a unit that's been mismanaged, that's a mess. Of course you can't work miracles on it. Manual release is part of intensive management, and intensive management starts when they take out that aerial photograph and put a line around a bunch of old growth and say, 'This is a unit.' That's when it has to start. They have to log it right, burn it right, plant it right, then brush it right, thin it right, you know, all the way down the line. They can't just wake up twenty years later and say, 'This is a mess. Go clear it up.' That doesn't work. Manual release has to be part of a whole system.

"This means that manual release is a silvicultural issue as well as an environmental issue. Let's say we woke up tomorrow and the EPA had just proved conclusively that herbicides were absolutely safe. Would that mean you should never use manual release? I say no. Why should you use manual release, when it costs at least somewhat more than herbicides? The biggest reason is because it's site-specific forestry. At the exact time and place of application, you can make a specific decision about what is best for that site. When they take a unit and spray it, there's no decision. The decision is based on a computer, saying this plantation is seven years old, and looking at the photograph it's got some brush on it, so you spray the whole thing.

"You can get into self-defeating practices that way. For example, I was in a unit the other day that was sprayed last year. But there was never really much brush there. We could have manually released it for $25 an acre. And what was the result of spraying? For some reason this area had tons of cedar reproduction. Just *tons* of it — I've never seen anything like it. But all the cedars were dead. Now if I could have gone in there and manually released it, I could have saved the cedars. A lot of them were right on the road banks. They were tight little clumps, which is great for holding erosion on cut banks. Now they're all dead. So what was the result of spraying? They hurt themselves. Eventually they're going to have clogged up culverts from erosion. They're not going to have any cedars up there, and cedar's a valuable tree. So what's the value of spraying? It's so *arbitrary,* and sometimes it just doesn't make any sense.

"Groundwork [a research group specializing in reforestation work] has done a study of spray areas. They questioned the *necessity* of spraying in some areas, so they went in both before and after the area was sprayed. What they found was that the decisions to spray were based on convenience. The Forest Service found a helipad somewhere and picked out units that were convenient to *it,* whether or not those units really needed to be sprayed. And Groundwork found that most of the trees in the units did not really need to be released. They also found a certain amount of damage to the crop trees from the spraying.

"With manual release we come in and say, 'What does each microsite need?' We make a human decision. Let's say I'm in a slide, and there's lots of erosion going on. I look in there and I see all these alders and brush, and they're holding that slide together. There's a couple of little suppressed firs. I say, 'If I cut those alders, that soil comes down and there's no fir.' So I don't cut them. I tell the inspector, 'Look, I didn't cut those trees because you need them there.' That type of situation repeats itself over and over in different sets of circumstances, where the *worker* can make a decision based on a microsite observation at the exact time of application. That's very important. That's good forestry.

"In site-specific forestry, instead of releasing *every* tree in the unit, you release the trees that you decide will be most commercially valuable. Therefore you don't use strict spacing, you go for reasonable spacing. You choose the healthiest trees. Don't say, 'We have 1,000 acres, and we want 520 trees an acre, and therefore the spacing is ten-by-ten, and therefore you release *every* tree ten-by-ten. . . .' That's how the big companies work. It's strictly geometrical. Instead, you say, 'Is that the most productive way to do it? No! The most productive way is to make the decision when we get there.'

"In site-specific forestry, you're on a south slope, 45°. The main reason to release is sunlight, so why should you release the north side of the tree more than a foot? There's absolutely no way you can shade the tree from the north side. On the other hand, the more you release downhill, the more sun you get in the early morning and late afternoon. If we ask the silviculturalist to put something like that in our contract, he just says, 'We have a problem if we put anything in our contracts that's a little complicated, because the workers don't understand it. Then the contractors complain that it's too hard.' That's bullshit. Everybody knows where the sun is, what the directions are. Everybody knows where north is. But they want to have a uniform standard and plug the worker into it.

"The workers can think, but they don't trust the worker to think. Give the workers any leeway and they'll do the easiest thing. The workers are stupid and lazy, right? Wrong. Forestry has changed. The workers aren't as transient as they used to be. You've got a lot of college education out there. You've got a lot of interest in forestry. That's a big change. And the old guard in the Forest Service doesn't like that. They can't deal with you the way they dealt with the bums. You question, you demand, you're a thinking human being. They're not ready for it. They have an image that the tree planter lives in cheap hotels, goes on long drunks between contracts, and cashes his paycheck at the bar. Whatever you tell him to do, he does.

"This leads into the labor issue. What's happening is that they're trying to reduce the number of people working in the woods. It's automation, basically. But what *we* think is that good forestry is people going out and making decisions on a real small scale. What *they* think, apparently, is that good forestry is very few people making broad decisions based on vicarious or indirect observations — aerial photography, computer projections, and so on. You get the impression that they would like each district office to have one forester, a room full of computers, and twenty secretaries to do all the bureaucratic paperwork. The forester just sits there and presses buttons, and that's supposed to be forestry.

"What we're saying is that forestry is thousands of people working in the woods and making money. That's good for the economy, and that's good for the forest. You have people out there looking, seeing, thinking, talking, communicating. Maybe they don't want people to *see* what's going on. That's very basic to the Forest Service and industrial foresters' opposition to this whole movement. They're very strong on the division of labor. They don't like to see workers integrated into decision-making positions. There's something about that that bothers them. But that's what we're pushing for. Workers *can* do research and make meaningful decisions. We're trying to develop a relationship between the worker-laborer and the scientist or technologist. It should all be integrated. That way you have a built-in interest in your work.

"You've got to have some interest in your work just to survive. Otherwise it'll wear you down, because it's not easy work. You get on the job at dawn, really early. You get out and you pack a gallon of gas and oil. You use a good, light saw. A heavy saw would kill you in this work, because you have to cut twenty to fifty thousand stems an acre. Your saw is always on full. You go down to the unit and you work six hours straight and then leave — no lunch or anything.

"All the units are so different. In one case you're surrounded by alder whips, thousands of them. You bend over and just mow them down. Then you encounter a wall of salmonberry; blackberry trailers are around your neck. You start reaching back with your saw to release yourself. It's like Vietnam. You get all wrapped up in brush and you try to cut yourself out.

"But it's spotty. A lot of these units include parts where there's nothing to do. That's when you make your money. But sometimes it gets really thick. The Coast Range is tough to work in because it's so steep, so incised. Every unit is just ravine after ravine, so the surface area of the unit is twice as much as the actual linear acreage.

"It's not easy work. And it's dangerous because you're cutting off all those little, short sticks, and when you fall down on them they stab you. The other dangerous thing is you're always getting whipped in the face and strangled by the brush. But it's not really serious. Mostly you get slapped in the eye or whipped in the lips and the nose. If you shut off your saw and start listening to your friends, you hear guys cursing and screaming. But you know they're not hurt. You know they're just getting whipped.

"It's not like tree planting. You can't talk when you work. It's too loud. You wear earplugs. If you want to take a break or talk to everybody, you have to wait until their saws idle down, and then rev your saw five times and maybe they can hear you. Then everybody takes their earplugs out.

"It takes a very special person to do this for more than five to ten years. It's worn my body down physically, very definitely. Real chronic problems: my back, wrists and elbows, shoulders. You find that with a lot of tree planters. Bad backs from bending down all day, tendinitis in the wrists and elbows from the shock of hitting your arms on the hard ground. For brush cutting, the most common thing is 'white finger' from the vibration of the saw. Like I said, it's not always pleasant. You're out there working in the cold rain. Your joints ache. You resent the government people who sit around in an office drinking coffee all day.

"I don't want to do this work forever. I don't *like* it that much. The best part of it for me is when I'm on that bus leaving the job. On the other hand, when you're in the age group of eighteen to thirty there are a lot of unemployed people. It's good experience; I've learned a lot from it. I've also learned a lot from being in a co-op. I've learned a lot from being in the woods. I've learned a lot about myself physically. You love it and you hate it. It's what we do. If I don't want to do it anymore, I'll change that. But right now that's what I do."

The critics of manual release often wonder why people would actually seek out such work. Why do it yourself when machines and chemicals can do it for you? Why turn back the clock of progress? Working in the woods is always grueling, but at least when you're logging you get the satisfaction of hauling out some impressive timber, which can be put to good use. Why would anybody voluntarily submit to working all day in troublesome, scraggly *brush*?

The very word "brush" is pejorative. Dow Chemical, attempting to encourage and capitalize on this negative image, advertises its herbicides by depicting dragonlike monsters hovering over the woods; these dragons, according to Dow, represent "The Brush Demon," which must be "tamed" with chemicals. Indeed, for many foresters the mere *presence* of brush constitutes a sufficient justification for chemical treatment, whether or not the brush is significantly impairing the growth of crop trees. In deciding whether an area is to be treated, foresters tend to ask two basic but superficial questions: (1) Is there any brush? (2) Can it be economically reached by helicopter? There are rarely (if ever) any site-specific inventories to demonstrate how each tree will actually be aided by the herbicides.

The Groundwork study of over sixty units slated for herbicide spraying within the Willamette National Forest revealed that all units were clustered closely around a handful of helicopter landings.[3] All but four of the units were within one mile of a helipad, and the remaining four were within two miles of a helipad. Since the economics of helicopter spraying involves high expenditures for fuel, equipment, and chemical handling, large acreages have to be sprayed in order to lower the per-acre costs. Consequently, every unit that appeared even moderately brushy and was within striking distance of a helipad was sprayed. Units in need of management yet not proximate to a helipad were neglected. In some cases, the units were too good to be sprayed: the crop trees were already topping out the brush. In other cases, the units were too bad to be sprayed: even if the brush were set back, there was not enough stocking underneath to make any difference. In all cases, crop trees that were being mechanically impeded by brush received no help whatsoever, because the brush was still left standing, even if the leaves had died off.

Groundwork's conclusion is that herbicides are not an effective silvicultural tool in most instances, whether or not they are hazardous to human health. The wanton use of phenoxy herbicides is only a symptom of poor management; the underlying disease is the use of area-wide techniques that ignore the idiosyncrasies of separate and distinct sites within the forest. As long as we apply area-wide treatments, we are bound to do harm as well as good. Only in site-specific forestry can we guarantee that each individual spot receives the treatment it really needs.

ONE MAN'S MEAT: WOOD WASTES

Site-specific forestry also leads to a more complete utilization of forest resources. A fallen conifer or a straight-grained hardwood is less likely to get lost in the shuffle if people are actually out in the woods instead of flying overhead. While people like Rick Kovens devote their energies to the perfection of silvicultural techniques, other holistic foresters focus their attention on transforming the wasteful attitudes and practices that have long characterized American forestry.

JIM DE MULLING
TREE FALLER AND ORGANIZER OF THE FOREST LANDS
AND PRODUCTS COOPERATIVE

"I was born and raised on a farm in Wisconsin. My father died when I was quite young, so we had to live with my grandparents. That's where I learned how to swing an ax, use a saw, and drive a team, and where I learned what wood was for. That was back in the thirties. Even as a kid I had jobs cutting wood, carrying it in. So I got to use an ax and a saw very young.

"I also got to watch trees grow when I was young. When I started high school, I joined a junior forestry club, because at that time in Wisconsin the two big mills were just on the verge of shutting down. Everybody was talking about what was going to happen two years from now, five years from now. What were people going to do? The school was in tune with the times, so they had a forestry club, and they had been planting trees for years, before the C.C.'s. So as a kid of thirteen, I got to start planting trees and learn what it was all about.

"On my grandfather's farm I planted some trees, red pine. It was pastureland, just rocks, nothing else. It was not much of a pasture anyway. So I planted them there, and they grew, and everybody driving along could see that they grew. When I was back there last time, I saw those trees. It's quite remarkable. Some of them have already been harvested. For once in my life I did something worthwhile. For young people, it's hard to imagine the trees they plant growing up and getting big and getting valuable and useful; but having seen it for myself, I know.

"Perhaps because I have worked around wood for so much of my lifetime, I value wood for what it can be used for. When I see the waste or the poor logging practices that result in trees being broken, I get very upset. There are ways of doing things properly to avoid the waste, to get better utilization. When I fall a tree, I know the tree should be felled to do as little damage as possible to the standing trees, and so that the tree itself will not break up too badly. Some you can't avoid breaking; in that case, you try to break them clean. And you try to lay them so it will be easier for the skidding crew to get the tree out. There's no use putting a tree in a place where they can't get it out.

"There's so much going to waste, and so many people on welfare, people who can't find a job. It doesn't make sense. Why can't these unemployed people do something with this stuff going to waste? Years ago I talked about it, hoping that somebody would take the ball and run with it. *I* didn't want to do it, because I don't have the time and I'm no organizer. I took it to the Board of Supervisors, and they referred it to their Forestry Committee. The Forestry Committee said it couldn't be done; there was no market for all that waste or somebody would already be selling it.

"Finally at the Community Congress some people got interested in it. We started having public meetings. We had a lot of meetings and did a lot of talking and we finally got our Articles of Incorporation written up.

"We have a few projects going now. We're doing some salvaging: posts, shingle bolts, lumber. And thinning some hardwoods out at the same time. The people doing the work are getting some experience. They're getting to know something about it. We have people planting trees. We had a crew out on Tostens' last year. I just went back through there and I couldn't find one dead tree, so obviously we did a good job. We *have* to do a good job. We have to do everything right.

"We're trying to develop all sorts of uses for this wasted wood: firewood, lumber, furniture making. A lot of this stuff is odd and unusual, and that's gaining more value every year. We haven't even scratched the surface of what we could be doing. Take tan oak, for instance. First of all, it could be made into lumber and furniture. It could be used for pallet stock. Almost any item that is made, like stoves or refrigerators, is shipped in some kind of container. For a rough box, tan oak would serve the purpose. What isn't used that way could be sorted so the high grade goes for paper, another grade goes for pressboard, another grade goes for something else. It could be made into fuel logs. The leaves and whatnot could be used for making a mulch.

"Almost any hardwood can be used if you treat it right. From the standpoint of beauty and workability, the California laurel is the best, what they call pepperwood here or myrtlewood up in Oregon. Chinquapin is nice, too. It resembles chestnut. It's a straight-grained wood, and it's fine for furniture or any kind of cabinet work. Another good wood for furniture or cabinetry — and it's unused in this country, to a great extent — is Pacific yew. But nobody's putting these hardwoods to use down here. There's a company in Oregon that sells golden chinquapin for $7 a board foot. Down here, those that have it either let it rot or else burn it up in their fireplaces.

"The thing is, there's no need to buy a sixteen-foot board to make a table four feet long. People think that if you can't get a long, straight log out of a tree, then it's not worth making it into

lumber. My contention is that even the short logs can be mar-
keted — if only you can find the market. There's always somebody
who knows somebody who knows somebody else who can tell you
where that market is. It's just a matter of getting together. That's
why we organized FLAPCO. We recognized that the forest is still
important to this area, that it has to be protected, improved, and
utilized. To utilize it better, we have to all work together. Anybody who's connected with wood in this area can be a member: a
landowner, a worker in the woods, a cabinet maker. It's not just
a workers' cooperative like they have in other places. An absentee
landowner can be a member, too. That way if a tree falls down
on his land, there's somebody around who can help him utilize it.

"Up in Oregon, it's mostly government land. They give out lots
of contracts, so there's lots of work for the Hoedads and other
worker collectives up there. Down here [California] it's different.
We're a long way from Six Rivers National Forest. We might get
a few contracts, but we don't depend on it. When you bid on a
government contract, very often you won't be the low bidder, so
you won't get the contract. If you get it, fine — but don't depend
on it.

"In this area, it's mostly private landowners. We're trying to
work directly with the people who own the land, because they
have the resources. We don't want to antagonize them. We're
actually trying to *create* the work by talking to the landowners
about how they can utilize their resources. We offer to buy their
salvage, the stuff that nobody else wants. They should be willing
to sell that to us for a very reasonable price, since nobody else
wants it. But we don't have much capital, so we give them a little
cash and then some services and then some stock. A person might
get enough cash, say, to pay taxes and deposit some money in his
co-op account. It's *his* money, but it stays in the revolving fund
for a year, so it's capital for us to use. Part of that money would
then go into reforestation on his own land, and that means jobs
for other people. The county gains because they get tax money
and they get people off welfare. They cut down the fire hazard.
They get reforestation on the hillsides, which holds back erosion.
A lot of places in this county are badly in need of reforestation.

"Somebody once said that the forest is the way out of economic bankruptcy. I agree with that. Not everyone can be employed as a timber faller on seven-, eight-, ten-foot trees. But there are plenty of other jobs to be done in developing the resources that we have all around us. If we were to manage our forested land in a manner similar to Switzerland or Sweden or some other enlightened country, we could have more timber growing, more people working, and everybody would be better off."

Today, it's not just a handful of idealistic conservationists who are seeking means to eliminate waste. The forest products industry, confronted with a dwindling resource base, is making major advances in residue utilization. Gone are the days of the old tipi burners, which used to send the mill wastes up in flames. Scrap wood, sawdust, and bark, once regarded as a nuisance, are now being turned into useful products. Mill ends are pressed, glued, and transformed into particle board. Residue from both the forests and the mills is chipped and turned into pulp for paper products. Sawdust is made into mulch or combined with other organic wastes in the manufacture of fertilizers. Even bark is now being utilized to make wax and extender for plywood glues and plastic moulding compounds by a few pioneering firms in the forest products industry.

Most significantly, wood wastes in all forms can be turned into energy. Since timber, unlike oil or coal, is a renewable resource, more and more attention is being devoted to wood as fuel. Wood can be burned in fireplaces and stoves for home heating. Charcoal can be burned in conjunction with coal for many industrial uses. Pressed logs made from mill wastes can be burned for either household or industrial use. Wood chips and sawdust can be burned in boilers to generate steam for heating, industrial power, or electricity. Finally, alcohol fuels derived from wood — methanol and ethanol — could theoretically be used to power cars.

These uses for wood energy are expanding day by day. Many large mills already meet most of their energy needs with their own "hogged" fuel made from sawdust or wood trimmings. The Pacific Lumber Company heats all the residences in the mill town of Scotia, California, by burning its wood wastes in a steam boiler. Throughout the Northwest, hog fuel is so cheap and plentiful that it provides the energy needs of many public institutions. In Grand Marais, Minnesota, the school system converted from fuel oil to wood chips — and they cut their energy-consumption costs by 75%.[4] In Burlington, Vermont, electrical production costs have been cut by one-third by converting to wood-chip generation.[5] As fossil-fuel costs continue to rise, the economics of wood-burning systems will become more and more attractive. Within the next century, wood may well become our major energy source, just as it was a century past.

What will be the impact of renewed interest in wood energy on the forests themselves? Possibly, the demand for biomass will be so great that forests which might have produced quality saw-timber will be converted to the production of fuel. The biomass boom could create an additional strain on the existing resource base. Already, we want more timber than the forests can provide. What will happen when we call on trees to provide us with energy as well as shelter?

If traditional patterns of boom-and-bust are repeated, the energy crisis will probably hasten the process of deforestation that we have been trying to hold in check. Yet, ironically, the use of wood for fuel could also provide the economic incentive for practicing a more well-balanced forestry. In the words of Norval Morey, one of the country's foremost advocates of wood energy, "Our problem has been in this country that there's never been a market for this low-quality wood. Every logger goes into the forest wanting to take the best logs, because he's either making lumber or paper or poles. To manage the forest properly, we need a market for this low-grade wood, and this energy is ideal for it."[6]

If the demand for energy becomes great enough, the raising of rapid-growth trees such as red alder may suddenly become an attractive prospect, and tree crop rotation may become economically feasible. The wastefulness inherent in the use of herbicides becomes more obvious as our need to utilize the biomass increases. Since a greater variety of tree species can be used to provide fuel than to provide lumber, there will be less of an impetus to reduce silviculture to monoculture. And the more diversity we can permit in the forests, the more opportunities we have for the practice of holistic land management.

The future, it would seem, augurs well for holistic forestry. It is a system of beliefs based on the principles that everything in our world is connected, and that, in this finite, interconnected space, nothing can be wasted or ignored. Current trends indicate that these principles will prove increasingly important as world resources grow scarce and as world politics grows increasingly complex. The fundamental techniques of holistic forestry are as simple as a Zen precept: use manual labor whenever possible; avoid artificial substances; treat each site as a specific problem in the context of a totally connected world ecosystem; find strength in balance and diversity; and — the golden rule — complement, rather than contradict, nature's actions as much as possible.

Significantly, holistic forestry differs from mainstream environmentalism and other voices opposed to industrial tree farming in that it is an active, not a passive, approach. Holistic forestry acknowledges the economic argument, affirming our rights and needs to employ our people and use the forest as a resource. As the energy crisis and other resource-oriented problems develop, passivity will no longer be tolerated with respect to the bulk of our productive forest land. It is the activism of holistic forestry that is its greatest hope.

Holistic forestry is still in its infancy. So far, it is only an idea, a direction, a consciousness. It postulates a middle road between overmanagement and no management at all. The exact route which that middle road will take has not been determined. The basic ideas are sound. But where do we go from there? How do we turn this idea into reality?

Part II

Hands in the forest: tools and techniques

Chapter 5

Harvest technology

Logging, the actual harvesting of timber, is the most dramatic and directly destructive aspect of forestry. The removal of giant trees is bound to have an impact on a forest environment, but that impact can be mitigated or increased by the kind of tools used to harvest timber. There is more than one way to skin a cat, and there are many ways, from horses to helicopters, to harvest a tree. Holistic forestry requires a knowledge of *every* available piece of logging technology, so the forester can select the right tool for the right job.

BIG CATS IN THE FOREST

Ever since it made its debut in the second quarter of the twentieth century, the caterpillar logging tractor has dominated the woods. Cats make the roads that crisscross the forests and offer ready access to the trees that will be removed. Cats prepare the beds upon which the larger trees will be laid to rest and help control the direction of falling timber. The same machines then hitch up the logs and snake them out of the woods to the truck landing. Cats even clean up after themselves, pushing around the slash, piling up the brush to be burned or churning it into small pieces that will be left to rot on the ground.

With power to spare, the logging tractor can alter the very face of the earth. In the past, natural barriers had to be avoided; today, they are simply removed. In the old days, the falling beds had to be prepared with piles of brush; the earth itself remained where it was. The old-timers clearcut and burned, but they lacked the wherewithal to completely change the face of the forest soil. When they built a road, they used their own muscle and sweat and the power of beasts of burden. Today the woods are honeycombed with roads. It is not uncommon for half the available space in a forest to be occupied by roads and skid trails, and for three-quarters of the earth to be disturbed by the movements of the ubiquitous cats.

A logging tractor disturbs the earth not only with its blade, but also with its crawlers, the ribbed tracks that grip the ground and give the cat its phenomenal traction and stability. Working under the weight of several tons, the ribs dig into the earth and churn up the top few inches of dirt, while the mass of the machine compresses and compacts the soil that is not overturned. Hoping to minimize damage to the soil while simultaneously increasing the speed of these mobile logging machines, operators in recent years have begun to switch from ribbed tracks to rubber-tired wheels.

Wheeled skidders have been heralded as environmental saviors, promising to relieve the topsoil of the adverse impacts of crawlers. They have, in fact, disturbed less earth and compacted the soil less severely, since they are generally not as heavy as the crawler tractors. But wheeled skidders have their own set of drawbacks. For one thing, they are far less stable. Many an operator has had a limb crushed under the weight of an overturned skidder. Lacking the power and traction of crawler tractors, they are not as useful for road building. Lacking the mass of crawler tractors, they do not serve well as anchors to control the direction of falling timber. Ironically, they actually do more damage to residual trees during logging operations, for the logs are harder to control when skidders turn sharp corners.

Wheeled skidders might be modifying logging practices, but they are hardly revolutionizing work in the woods the way caterpillar tractors did a half-century ago. The cats opened up hitherto inaccessible regions, paving the way for a brief, short-sighted, and unplanned plunder of the timber resources of the backwoods. The adverse impacts on the forests were profound, but the social impacts of cat logging have been equally profound, and these are rarely talked about or even noticed. Cats have granted instant power and liberation to the working loggers.

In the old days of steam donkey logging, the workers had to be organized in large crews. The life and livelihood of each man depended on his co-workers and on the company. The fate of a worker might have been determined by a donkey puncher a half-mile away. If the whistle punk signaled a second too early, a lumberjack might have found himself cut by a cable, crushed by a log, or suspended in thin air. No woodsman was boss for himself; even the foreman had to answer to the company, which was run by capitalists in faraway places. But the cat has changed all that. Today, the single-man logging show is a living reality. A cat-skinner can fall and yard his own trees with no help from anyone else. Throughout the backcountry, small parcels are logged by individuals or small groups of men who are not dependent on others for their own safety. Even on large parcels owned by corporations or the government, much of the work is contracted out to small logging crews and independent operators. Land ownership has become increasingly centralized, but the existence of the caterpillar tractor has kept the actual work in the woods from becoming equally centralized. The owner of a cat has power over the trees and power over himself. He can contract out to work for others, but once in the woods, his world is simplified into three basic elements: the logs, the logger, and the logging tractor.

JOE MILLER
CATSKINNER

"I was a mechanic in the [San Francisco] Bay Area, fixing cars for other people, fixing other people's problems. I wanted to leave the city, so my wife and I came up here. Our immediate reaction was to find a piece of ground to homestead. We walked around and found one with water and a lot of wood. We didn't own the land; we were just squatting. We proceeded to build a house out of a redwood tree that had fallen down and shattered. I saw what I could do with that one log, and I saw how many logs were lying around. I was impressed.

"We were living near the Commune, and they had just accepted the Lord. So every time we came out of the mountains to be with people, we heard talk about Jesus. It started making things fall into place. There was no pressure, no shoving anything down our throats; so we accepted the Lord.

"I never lost my connection to the wood that was all around me. I started making fence posts and grape stakes. I started making shake bolts and taking them into town. Everywhere there were these big old chunks of logs laying around, and that was money.

"I started working on a truck. It had a winch to drag the logs out of the brush and load them on the back of the truck. Then I'd drive them down to where they could be made into fence posts, shake bolts, or whatever.

"By that time we were living at the Commune, and as it grew I had less and less time to work on my own ideas. I was working more toward the benefit of the community, working on people's cars and fixing washing machines. It got to the point where I had to decide whether I was going to live my life full-time for Jesus, administering His gospel, going out to the four corners of the world. But I had this intense desire to go get those *redwood logs*. It sounds corny, I know, but they held a special attraction for me. It wasn't like working on cars, it was like panning gold. There the logs were, and people were just ignoring them. But I knew what I could do with them.

"So after living at the Commune for about two years, we left, borrowed some money from my mom, and bought a four-wheel-drive truck. It was called 'The Frankenstein.'

"My tools were a chain saw, a hammer, wedges, blocks, and a piece of cable. I'd go out and cut the log to a size I could pull, and get it out in the road and make fence posts or grape stakes or whatever I could make out of it.

"I started hustling landowners to work their wood. Then I got my dad to put up the money to buy a piece of land. I got a World War II half-track with a winch on it, which had more traction than the four-wheel drive. And I got another chain saw.

"And then I got my first cat. Sold the half-track, sold the truck. The cat was really just the thing. It didn't slide. You didn't get stuck. You could turn on a dime. You could pull anything you wanted—and it didn't break down. It was just a hunk of iron made to order for pulling logs out of the brush.

"It was a really old cat. If I didn't keep it loaded with Power Punch, it burned eight gallons of oil a day. But I kept it so glued together with Power Punch, which is like STP, that I got it down to two or three gallons a day.

"Every time I looked around there was more and more wood laying around, more wood than I could possibly work. So I discovered how to send logs to the mill instead of working them myself. I had to go to Eureka to pick up some machinery I had up there, and I rolled a few small logs on the truck and took them to the stud mill. I was very upset that they didn't pay me when I delivered them, because I was living hand-to-mouth. But when I got the paycheck, I was thoroughly shocked, because it was so high. It turned out I made more money with less work. So I started hustling as many logs as I could to the mill.

"About that time I started doing "green" logging. I figured that if there was that much money in sending the logs to the mill, I might as well start selling green logs, so I could pay my property off. So I started falling trees, and I had to buy a loader to put the logs on the big log trucks.

"I didn't have anybody working with me. I'd be jumping in and out of the cat, which is some of the hardest work there is. On the hillsides it's so steep and it's so high up to the seat that you find yourself climbing around like a monkey. You park the cat, take the brake off the winch, and pull out this one-inch winch line. If the log is downhill, you can lean into the cable and turn the winch, but if the log is uphill — well, then you really have some work to do. You hook the line up to the choker, which is hooked up to the log, and then you go back up to the cat and put that winch in gear and bring it in. Then you build your turn of logs, hook them together, and take them down the road to the landing. All this time you're getting on and off the cat, climbing up to the seat, and jumping down again.

"So here I am green logging, doing it all by myself, and finding it's very slow. So I started thinking about employees; but employees meant ties with the government, insurance, and responsibilities. I didn't like that, but I did it anyway. I went out and got workmen's compensation and filed with the state and the federal government.

"The first guy that worked for me was a good worker, and I enjoyed him, but he wasn't always there. He lived off in the woods and didn't have a phone. A lot of times he just didn't show up for one reason or another, and there was no way to reach him. Since then I've had sixteen employees, four or five at a time. But it just hasn't worked out. Motivation is the biggest problem. Motivation to *come* to work, motivation to *work,* motivation to do *good* work. And motivation to enjoy one's existence.

"From now on, I intend to work mostly by myself. I want to work with my ideas through my own hands. I don't want to be pushing my ideas through someone else's hands, because that doesn't work. I can't seem to motivate anybody to do it the way I would do it. And I get more enjoyment out of doing it than getting someone else to do it.

"My machines are my best employees. Especially my cats. I can manipulate them. I can have them do whatever I imagine them to do. I can run that machine to its maximum. I know how it runs. I know how much power I can use. I know when it needs grease, what parts are moving freely and what parts are sticking.

"The machine becomes an extension of your hand. All the time you're seeking its limits. Of course machines break down if you take your mind off what you're doing. When you're sitting there grabbing the levers and cussing at it, it's going to break down. But if you're sitting there feeling and flowing with the machine, it doesn't break. It works for you, because it's just a machine. It's not out to get you.

"I love working on my machines. I have a shop and there's a proper tool for every purpose. If you don't have the proper tool, it's frustrating; but I know that, so I don't mess with it. I get the proper tool or figure out a proper substitute. There is a right way to do everything. If you sit there and think about it long enough, you'll figure out a solution. It's creative. You find out new ways of doing things, and when they work out real slick, they make you feel good at the end of the day. Sometimes the work takes a month or two, but then the finished product rolls out: slick paint job, brand new decals, and nobody else's Mickey Mouse fixit jobs in there. You've gone through and freed everything. You've cut loose all the things that were welded and you've fastened them properly with bolts or pins or whatever. All those sloppy joints, you've shimmed them up. And *you* feel shimmed up. You feel together.

"Caterpillar tractors are designed to run forever. They just don't wear out. You can take out the bolts and put them back in every ten years for a thousand years or whatever. Take my 1946 cat. Some guy must have known just exactly how much steel to put in everything. It's just a good tool. In my business, it's the best tool.

"I use my cats for everything. I even fall timber with a cat. If I need to knock out a high spot to make a bed, the cat will do it. If I need to pull the tree, the cat will do that, too. If you run into any problems, you've got a cat there as one of your tools. With a cat you can line all the trees up in a straight line, so you don't have to sweep the ground with your logs. You can fall the tree any way you want to. You cut your tree and leave your hinge, but you make that hinge extra thick. You set your wedges in the back cut, then climb in the cat and pull it over center and there it goes. You get back out, buck it, limb it, and choke it. Then you drag it out with the cat. The cat does it all: it builds the roads, falls the trees, and hauls out the logs. It's the greatest tool I can imagine."

One man's virtue is another man's vice, and the logging tractor has led to rags as well as riches. The cat's power and versatility have enabled it to do the job of many men — and simultaneously put workers out of their jobs. Joe Miller has found he likes to do the whole show himself, but not everyone has $150,000 to invest in a new logging tractor or the ability to make an old one run like new. For lumberjacks who are not their own bosses, the caterpillar has cut a road directly to the unemployment office.

If used correctly on gentle terrain, the caterpillar tractor can be an invaluable tool. But the cats have been overused. Their power is just too great — and their abuse all too common. Hills are pushed around by catskinners like sand in a two-year-old's sandbox. The wounds to the earth are not consciously inflicted, but when such enormous power is at your command, it is hard not to use it.

HORSE POWER

What if the cats did not have their destructive powers? What if these mobile yarding machines could simply pick up their tracks and step over the small irregularities in the earth's surface, instead of driving relentlessly over every little obstacle? Indeed, what if they made logging tractors out of flesh and blood instead of steel?

JOAN AND JOAQUIN BAMFORD

HORSE LOGGERS

Joaquin: "I started out as a choker setter for Weyerhaeuser back when I was in high school — 1959, I guess. Then I worked up to rigging slinger, and then I was a hooktender. Then I ran cat and ran a skidder. Then I worked on a yarder. And I worked as a foreman. I've done many things since then, but primarily everything has been in the logging business.

"Riding back from work a few years ago, this guy says, 'You know, we ought to get out of this business.' I said, 'Yeah, I suppose we should.' I wasn't happy with the outfit I was working for. He said, 'You ever thought about horse logging?' I said, 'No, not really.' He said, 'Well, we should look into that.'

"I've always believed that if you want to look into something, you go to whoever knows most about it. I heard about this fellow Al Merrill. He was about sixty-five at the time, and he was still out there logging. I went out to meet him and went around with him, helping him out when I could, watching him and asking him questions. Of course, the first thing he said to me was: 'If you want to go into horse logging, you might as well figure on starving to death.'

"So we set out to get our finances all squared away before we started. You don't want a lot of payments to make if you're in horse logging. You need to have a little bit of money in the bank. And you need to have an awful lot of experience behind you, not only in logging, but in horse psychology, horse sense. What to do when a situation arises. How to move. How to be light on your feet. You have to be able to perceive what's going to happen to you fifteen, twenty feet down the road. You have to be able to jump out of the way, stop them, turn them. Think what the horses are thinking: whether they are spooked or scared; whether they'll stand for you; whether they'll back up for you if you want them to. If somebody else is out there, make sure you're never in a bind in case they happen to say a word that means for the horses to move ahead or move back.

"A workhorse adapts easily to logging. If he's used to pulling things, it only takes an hour or two to get used to working in the brush. But adapting to the person is the important thing. Bill [one of their horses] wouldn't even go into the truck until he had confidence in Joan. But he sure has confidence in her now."

Joan: "Bill's never refused to pull for me. When he gets excited, he'll go uphill pulling these great big ones sometimes, and he'll be digging in the ground with all fours, and you can just see his muscles working. But you should never put your horse on something he can't pull. That really ruins a horse. Makes him balky.

"Every time I take a log to the landing, I always let Bill rest. Because I really appreciate what he's doing for me. He enjoys the work. I enjoy the work. It's not for the volume that we do it. It's a challenge. Every log you take down is a challenge. You have to figure out the angle, if it'll bang up another tree, if it'll get you as you go around the corner. There are so many things. Every time you get down to that landing, you feel you've really accomplished something. Then you come up and you take the next one out. It's always exciting. It's never dull. But it's so quiet, so peaceful. And it's so nice to be working with Bill.

"I've tried driving a team, but they were too fast. . . ."

Joaquin: "At five-foot two, you know, she couldn't see around two rumps. She was running between horses, trying to see over them or around them."

Joan: "I'd drive them into a tree. There would be a tree between the harness and everything. One horse, you can angle him to where you see where you're going. I don't think I'd want a team now. Not after pulling with Bill."

Joaquin: "But you can't log as big a log. I mean, with a team you can log some really big logs. Most all the horse loggers I know have two horses. Not that they always use the two horses, because when you're thinning small timber, it's better to use one horse. You can get between the trees better, and you don't have to make a real wide skid trail.

"When I was horse logging with a team, I did a fifty-fifty split with the landowner and I came out with a couple hundred dollars a load. You can figure three, four loads a week, so you can get maybe $600 a week — if you're into the right type of timber and if you're logging steady.

"So you're looking at $2,400 a month. It depends on your job. And you need more than just one job going. You've got to be able to work at several different jobs, so that if a week of steady rain comes along, and you can't work with the horses, you can go work ahead, falling your trees. And it depends on your team. You have to know your team."

Joan: "Joaquin had a neat relationship with his team. A lot of horses won't stand in the brush, but he had them so well trained, they would stand there all day. If a tree got hung up, the horses would pull the tree down for him."

Joaquin: "If you get a team that's easy and slow and pays attention and is real good at it, then you're going to be a success at horse logging. If you get a team that wants to run, that's a little

Modern horse logging

bit jittery, that jumps onto the turn or doesn't work together, then you're going to get yourself hurt. We've had several horses that we had to get rid of because they got nervous and excited and didn't pay attention to you.

"The thing about horses is you can't get in a hurry. No way. Whenever you start getting excited and worried about getting enough volume in, then something's going to happen. You can't hurry the horses. You just have to take one step at a time. It's not like machine logging. When you own machinery and you have big payments (I mean they don't *give* that machinery away), the logger has to rip and tear and go right after it. There is no being really careful. He's got to get in ten loads a day, or whatever it takes to make those payments.

"In horse logging you have to look at the timber before you can go and say, 'Yeah, I can log that.' The best type is second growth, timber that ranges from fifteen years to forty or fifty years. You want it small, but not too small. We usually won't go into timber that is less than twelve inches on the butt.

"And you need to look at the land. You want a downhill pull, if at all possible. You look at your terrain. You find out what the soil is like: whether it's loose soil, whether it's on a steep hillside, whether there will be erosion where you drag the timber down. You have to look at it as a conservation practice.

"When you start out on a job there's a lot of preplanning. You go in first and clear the brush out of the way. You have to take off the bottom limbs so the horse doesn't gouge himself or hurt himself. You cut all the vine maples and the underbrush down low enough so when the horse steps on them he's not going to hurt his feet."

Joan: "Some people don't do all that. They let the horses work in the brush. They don't prepare the way for them. They leave limbs sticking out. Those horse loggers often have big scars on the horses' legs. It's a lot of work to clear the way, but we're doing it for the horse. We don't want to hurt a fine animal like that."

Joaquin: "About three years ago, a lot of individuals went into horse logging without the proper equipment or the proper horse. They had chains for tugs; or the collars didn't fit quite right; or they didn't have a pad. A lot of things were just thrown together.

You know, living off the land and having a horse and doing horse logging — boy, that's right back to nature. The people would go in and log properties where they really didn't have any experience in falling timber or working a team. They ruined a lot of teams and made a lot of people in the area upset with horse loggers."

Joan: "Remember the guy we saw who had a large saddle horse that he used as a draft horse? He wasn't built to move timber like that, but they used him."

Joaquin: "His bone structure wasn't really heavy enough; his muscle wasn't heavy enough; he didn't have the chest on him to do that type of work. He was a big horse, but he just wasn't a draft horse."

Joan: "The first thing you have to know is how to relate to animals. You have to realize what your animal is thinking at all times. People think they're dumb, but they know more than you give them credit for. In fact a lot of horses are more intelligent than the people I've seen trying to run them. It's amazing what a horse will do if you just get the communications squared away."

Joaquin: "You can get a bond with a horse. Like for lunch, while we have our peanut butter and jelly sandwiches, he gets his oats and grain. They really appreciate that. They appreciate having a blanket put on them after they're hot. They appreciate the hay they get. They appreciate getting some praise after they get down to the landing with a big log. They just eat that up."

Joan: "And if you're upset, you can transmit it to them."

Joaquin: "If you go out there and you're afraid, then that transmits to them. If you're not sure, then that makes them spooky. Because they know that you don't know. They can tell by your voice."

Joan: "Animals have their own personalities, just like people. Bill, he wants to go to work. He gets excited when he gets his harness up."

Joaquin: "When you get his harness, you can almost see him grinning."

Joan: "Because he knows he's going to be used. And then when you bring him back, he's really excited by that, too. You put him out in the pasture and he rolls and he just relaxes. He's glad to get home, but he's glad he worked."

Joaquin: "Again, you have to relate it to people. People want to be used. They want to do *something*. That's the way Bill is. He was bred to work. He doesn't want to be worthless.

"I'm the same way. I like to work. I love to fall timber. I make a real game out of it. I really enjoy pulling a tree and driving in a stake and trying to get it to go exactly where I want it to go. It's a real challenge. It's exciting to be able to fall timber where you want it — most of the time. If anybody tells you they never made a mistake falling timber, they're either lying to you or they haven't been in the brush very long.

"Falling timber is something you can do anytime. You don't have to wait for good weather. Like today it's raining, so this would be a good time to cut. Between falling timber and horse logging, you can keep working year-round. This time of year, horse logging will do a cleaner job than a cat. When the ground is wet and the cat gets into a pull, the cat tracks spin and immediately start to tear the soil up. Water will run down that and cause erosion. When you lock a track, it spins and compacts the soil to make a rut. With a horse you don't have that problem. Of course, a good catskinner working on dry soil doesn't do things like that."

Joan: "But the cat runs over the little baby trees, the new growth; and the cat can't get where a horse can go, between the trees."

Joaquin: "But normally when you do a thinning, you don't want the trees much closer than a cat track anyway. When you thin them out, you want to have ten, twelve feet between trees. If you have a proper stand, a cat can get in, get out, get around. But they do have a tendency to scar the trees up more, because they have such power. It depends on the individual. A good catskinner can do just about as good a job as a horse logger."

Joan: "In horse logging, you want to leave it as natural as possible."

Joaquin: "Ninety-five percent of everybody who's got timber, if they're going to do any logging at all, would like to see at least a couple of loads horse logged. So there's no problem getting jobs. But you have to convince the landowner that you're really professional. You have to be able to talk to them intelligently about conservation, about how you would do it. Take them out and say why you would take each tree: it's got a sucker knot in it, or there's too many limbs on it, or it's cutting out the growth of other trees. Build the owner's confidence in you. Once you start that kind of a program, it's easy to get work because word of mouth will spread in the area. People will want horse logging because it's interesting. They want to see what it's like. They want to see a horse work."

A horse doesn't weigh as much as a cat. It doesn't push around the topsoil. It doesn't spin its tracks, create erosion channels, or significantly compact the earth. It doesn't bump into residual trees. All in all, a horse is gentler on the forest than a logging tractor or a wheeled skidder. And a horse consumes renewable resources — hay and grain — rather than unrenewable fossil fuels. A horse is not made out of metals that must be extracted from the earth. It is not manufactured in factories that dump polluting by-products into the air and water. Environmentally speaking, a horse is preferable to a cat in every respect. And it is also cheaper to buy.

For thinning small timber or harvesting poles, a horse can hold its own; but try as they might, even the mightiest draft horses can't tug on a large log as well as a cat. Horse logging is slow and tedious work. When the timber is small and scattered throughout the forest, environmental benefits can offset the relative inefficiency of horse logging. But when the timber is large and most or all of the trees are to be harvested, the horse will be competing out of its league. Horse logging, in short, is a valuable tool, but it is not a panacea. Like any good tool, draft horses should be used only on the jobs for which they are best suited.

HIGH-LEADS AND SKYLINES: CABLE YARDING

A horse might be out of place on a clearcut of mature timber, but other alternatives to cat logging are specifically geared to such concentrated operations. Cable systems of various sorts have been modernized since the days of the old steam donkeys. Modern, gas-driven yarding machines with portable steel towers have replaced the donkey engines and the spar trees that once supported the cables. Today's portable yarders can be driven along

A spar tree of steel: the mobile yarder

logging roads, stopping at convenient locations to extend their steel cables thousands of yards into the woods. In the *high-lead* systems, the cable is hooked on to a log, suspending the front end while the tail end of the log drags and bounces along the ground. In *skyline* logging, a cable is suspended not only from the yarding tower, but also from a high point at the tail end of the system. A carriage runs along this elevated track, and a log attached to the carriage can be reeled in to the landing without ever touching the ground.

Cable logging has enjoyed a renaissance in recent years, largely because it is supposed to be gentler on the earth than cat logging. Since the cables commonly reach a quarter-mile or a half-mile into the woods, the need for logging roads is greatly reduced. And since the machinery is confined to the roads, there is less soil disturbance and less soil compaction. There are numerous studies

Skyline yarding with a grapple rig

that compare cat skidding with cable yarding, and they all point to the same obvious result: a logging tractor zigzagging through the woods will do more damage to the earth than a cable dragging a log (high-lead), and the least damage is done when both the cable and the log are elevated (skyline). One such study, for instance, revealed that a logging tractor severely disturbed 36.2% of the area, slightly disturbed 37.6%, and left only 26.2% totally undisturbed; a comparable skyline operation, however, severely disturbed only 2.8% of the area, slightly disturbed 22.4%, and left 74.8% totally undisturbed.[1]

Another virtue of cable systems is that they generally yard the logs uphill, whereas tractors function best when skidding the logs downhill. The landings for cable systems, therefore, tend to be situated on ridgetops, where they do little damage; tractor landings, however, are often near streambeds, where the displaced earth is easily washed into the major waterways. Logging trails that lead to a ridgetop landing actually tend to spread the flow of water in several directions, for the water is *diverging* when it follows the trails back downhill. By contrast, logging trails converging on a streamside landing tend to channel all runoff into one central area, intensifying the damage due to erosion.

Cable logging does have its drawbacks. The landing pads tend to be large, for they must accommodate not only the logs, but the yarding machine as well. When mobile yarders use the roads as landings, the roadbeds have to be significantly wider. The extent of soil disturbance increases with the *square* of the width of the roadbed, so even a slight increase in the size of logging roads is likely to take its toll. A study of several logging sites in California found that cable yarding was preferable to tractor skidding on gentle or moderate hillsides; but on steep slopes with a 70% grade or higher, "cable yarding produced from two to twenty times as much erosion as tractor operations."[2] Ironically, cable yarding is often used on steep or unstable hillsides precisely because it requires fewer roads and is alleged to produce less erosion, but it is on these very hillsides that the larger landings and haul roads will cause the most damage.

Cable yarding is also difficult to adapt to silvicultural systems other than clearcutting. When a portion of the overstory is to be retained in a selective or shelterwood harvest, it can prove troublesome to lay out a quarter-mile stretch of cable that will not run into residual timber. A cable logger who wants to avoid clearcutting must still provide pathways for his cable by removing intermittent bands of trees. This can be done in two ways. (1) The logger can keep his pathways narrow, while reaching sideways into the woods with special tag lines attached to the main cable. In this manner, he can reach almost any tree in the forest without removing its neighbor. This type of selective logging can be (and has been) done, but it requires a sophisticated cable system with a locking carriage on the mainline, as well as a supremely skilled operator. (2) The logger can harvest only those trees within easy reach of the main cable. In this type of "strip-cutting," long, parallel swaths of forest are removed, while the timber between the swaths is left intact. Strip-cutting is, in some ways, a viable alternative to clearcutting, but it is not truly "selective" forestry: the choice of residual trees must be based on an arbitrary geometric pattern, rather than biological or silvicultural criteria.

Then there is the question of cost. High-lead yarding runs about four or five dollars more per thousand board feet than tractor skidding; a skyline system generally costs about twice as much as logging by tractor. The extra cost stems from two factors. First, the capital expense is higher: a high-quality, versatile yarder costs several times as much as a logging tractor. The operating costs are also higher, for a cable system requires a larger crew. The cables must repeatedly be set and reset in new locations; this takes time, labor, and, therefore, money. In a skyline system, the time and energy required to reset the lines can be particularly formidable.

Of course the extra cost might be justified in some situations. An extra $5 per thousand board feet is not that significant when stumpage prices are over $200 per thousand. The increase could easily be absorbed by the finished product, as "the cost of doing business." But if a more expensive method is to be chosen over a cheaper one, there must be a good reason for it. A significant reduction in erosion is certainly a good enough reason. For logging shows with fortuitous ridgetop landings and haul roads, cable yarding might be ideal. In situations where the landings and haul roads must be located on steep sidehills, or in situations that call for selective or shelterwood harvesting, cable yarders can sometimes be as destructive as tractors.

UP, UP AND AWAY: BALLOON YARDING

Some skyline systems have been modified in recent years by the addition of giant helium balloons. Logging balloons measure over 100 feet in diameter and can provide the lift for payloads weighing over 20,000 pounds. The balloon is attached to a cable system directly over the carriage that transports the logs; since the pull of the balloon is straight upward, it suspends the logs in midair, while the yarder guides both the logs and the balloon gently along the cables to the landing. The balloon eliminates the need to suspend long cables from a series of spar trees, for the balloon itself functions as a mobile spar tree, with the suspension always centered directly over the logs. The logs, therefore, can be yarded without ever touching the ground, eliminating all soil disturbance on the slopes and protecting the logs themselves from damage by rocks, debris, or irregular terrain. And since the balloon can function on cable systems up to a mile in length, the haul roads and landings are fewer and farther between.

One major difference between balloon logging and conventional cable logging is the direction of the yarding. Most cable systems yard uphill, so they do not have to put a brake on the immense power and momentum of downward-moving logs. Balloon logging, however, functions best with downhill yarding. Braking the logs is no problem for a balloon system, since the upward pull of the helium counteracts the downward pull of the logs. When logs are yarded downhill by balloon, the vertical thrusts of the payload and the helium balance each other, so the yarder has only to provide enough energy for horizontal movement. On the haulback phase of the cycle, the balloon lifts itself effortlessly into place to receive the subsequent turn of logs. The energy requirements for downhill balloon logging are therefore minimal. The yarder uses only about one-fourth as much fuel as conventional cable methods, which must haul the logs uphill. But when the balloon itself tries to log uphill, it runs into a bizarre problem: the balloon can lift the logs to the landing with ease, but then the yarder must put out its maximum effort just to drag the upward-pulling balloon back downhill. The energy required for uphill balloon logging is actually more, not less, than that for conventional cable systems![3]

FAYE STEWART

BALLOON LOGGER

"Many years ago, we were having difficulty logging one area down on the coast. We were having lots of slides on midslope roads. The state of Oregon had laid these roads out, and we were forced to build the roads to their specifications. But we ran into difficulties. In fact, we destroyed two big ranches by the slides from the hillsides sliding down and almost killed some people.

"From that experience, I felt that there had to be a better way to log. I had logged with cats, with high-lead logging, and with skyline, but in that type of terrain, none of those was the answer. I really got interested in helicopter logging, but the expense was too great.

"A close friend of mine was with an engineering firm in Seattle, and he told me about the Swedes experimenting with balloon logging. Well, that aroused my curiosity, so I contacted the Swedes and got some copies of some of the literature that they had written about it. Actually, it wasn't very helpful, because they had just captured some old British barrage balloons and tried using them with little success.

"I contacted Goodyear Aerospace around 1960 and finally got a balloon in the air in 1962. It was a V-shaped, aerodynamic balloon. The first balloon flew well enough that we went ahead with the program and kept working with it for the next five years. But Goodyear was never able to get any stability in its balloon. It flew like a kite without a tail on it.

"I was about to give up on the program when Raven Industries came on the scene with a natural-shaped balloon that they wanted us to try out. It was a small one that looked like an ice cream cone. We had been told by all the balloon experts up to this time that a natural-shaped balloon would fall out of the air when you got winds over fifteen miles an hour. But we put it in the air and knew within a week's time that the balloonists and aerodynamics people were giving us a bunch of hot air, because that balloon really did a job flying.

"It took us two to three years to get our ground support equipment capable of handling these new balloons. We feel now that it's an economical unit, capable of doing almost all of the things that we originally expected out of it. Our ground support equipment is working well. We've made numerous changes in the balloon design, and we have a balloon now that is capable of flying in fairly high winds, capable of working in all sorts of weather. Fog and clouds don't bother us a bit. We *love* fog. We've gone for days and never seen the balloon. Snow is what will tie us down. Snow will pile up on the balloon, and it'll lose its static lift.

"Our old balloon had an area on top that was sixty feet in diameter and perfectly flat. When the snow would get on there, it'd start to dimple, and that'd be the biggest damn bathtub you've ever seen when the snow started to melt. We got caught up on Mt. Hood with one of these flat balloons, and we actually got up on it and shoveled. One night we had about three feet of snow, and we had a hell of a time getting that snow off. I said, 'There must be a better way to do it.' The next time we tried letting her up a little bit, and it started tipping the snow off the side.

"Our new balloons have a hemispherical top. We really don't know what snow is going to do with them. But if you had that balloon in the air 250 feet, I don't think snow would ever force it to the ground. I think it would just tip over and dump the snow first. But we don't like to leave the balloon in an area where we're going to have heavy snow. If we're working in the mountains, we have to move the balloon out of there during the winter. We'll have our main winter bedding area down at a lower elevation.

"When a windstorm comes up, we worry about getting a limb through the darned thing. We do our best to find a bedding area where the wind can't get directly at us. We keep a man at the balloon all the time, and we also keep sandbags in the area. When the wind starts blowing it around, you've got to keep the darned thing flat on the ground.

"We've had some freakish experiences with balloons in the

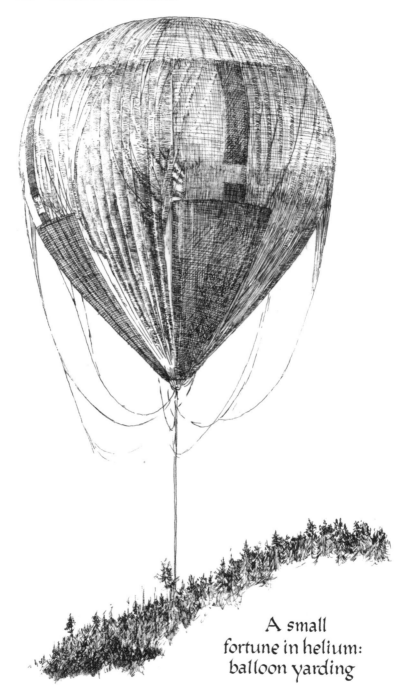

A small
fortune in helium:
balloon yarding

past. We had an ice storm catch one and bring it right down over the tops of some 150-foot fir trees. Of course it just knocked all the gas out of it. Fortunately there wasn't any high wind or it would have torn it apart. I sent a climber up and we cut as many limbs as we could. We cut all the chunks and logs out from in and around the balloon, patched the son-of-a-bitch up, and went back to logging with it in a couple weeks.

"It's surprising: it's not all that tough putting a balloon back together. A lot of people think when you lose a balloon that it's a catastrophe. Actually, it's not all that bad. You can just keep patching it up.

"With the new fabric we have, a balloon should last for at least five years, and probably near to seven. Even if you never have to patch it, you need to paint it. That's the main thing: keep your balloon painted, so ultraviolet rays are not raising hell with the fabric. This paint is specially designed to take care of that.

"To inflate the whole balloon costs over $25,000. That'll buy a good number of short beers; so you don't want to have to blow it up too often. Twenty-five thousand bucks — psssh — into thin air. The balloon we're flying now down on the coast has been inflated for the past four years. We use about 15 % of the volume during a full year; that's if you have good material, and if you keep it painted. You have to replace that much helium every year just for normal maintenance.

"If we run into any trouble, we'll bed the balloon down, but it stays inflated while we work on it. The only time we bed it down is if we want to paint it, inspect it, put more gas in it, or we know that a storm's coming in. We have a D-8 cat that we adapted especially for moving the balloon. We run that cat down the road to the bedding area, whether it's a quarter of a mile or five miles. If we're moving it any real distance, we'll just run that cat on a lowboy and have the truck haul it around with the balloon flying right over the top. Away we go. We've moved as much as ninety miles in one evening.

"So we can keep the balloon inflated, even from one unit to the next. We prefer to log clearcut units. It makes it a lot simpler. For some reason or other, we've recently been logging over reproduction. That involves concave slopes, so our lines will stay over the top and not do any damage. We've tried logging selection, but it's just no good.

"We logged a unit on the Cottage Grove district of the Umpqua, and they wanted to leave all of the old incense cedar standing. Somebody just had a brainstorm. I went up, I fought them, I did everything I could to get them to change the damn thing. Well, we left nothing but a hell of a bunch of snags. We left some of the trees in good shape, but an incense cedar, to start with, is the most brittle wood in the world. You hit a damn limb, and, hell, the limb is gone. We were logging a long way, almost 4,000 feet. You get any drift in that balloon, and it's going to get the cables on the wrong side of the tree. I wanted to cut corridors, but they wouldn't go for it. So now they have a damn snag patch, and it's a black eye to everybody.

"But for logging a clearcut, you can't beat a balloon. Anytime you're logging on sensitive soil, where you shouldn't drag logs, I feel that a balloon should be used, or a helicopter. When you're in rocky, rough terrain and dragging logs will damage that wood too much, or when you have an area that you can't reach with a conventional high-lead system without putting roads in that would do irreparable damage to the environment, why then I think one of the aerial systems should be used: skyline, helicopter, or balloon.

"Of course we believe in the balloon method of logging, but the helicopter is also a necessary logging tool in the western United States. Lots of things can be done with a helicopter that can't be done with a balloon, just as things can be done with a balloon that can't be done with skyline. Helicopters are very mobile. They can pick up a tree here and move over another hundred yards and pick up another old-growth tree. In other words, in overstory removal they do an excellent job. There are small patches of timber scattered all over roadless areas of the Pacific Northwest. The timber is still very valuable in these areas, and the helicopters can go in and get the timber out.

"With a balloon we need a road within about a mile. The second problem we have with the balloons is if there is high, young growth on a convex slope, our lines will slap the understory and damage it. On a concave slope we can log over it and pick the

trees out from the understory without any problem. But the helicopters can go into these convex slopes and pick up the overstory
material without hurting the understory. And they can fly materials up to three miles. It's really not very economical, but as
badly as the United States needs the wood fiber, helicopter logging can be worth it.

"With our balloon logging, anytime that we get out over 1,500
feet, we are more economical than the skyline, even on shows that
the skyline can work on. A skyline is a large line, it's hard to
handle. With balloon logging, our lines are light. We only use a
one-inch cable. In skylines they use up to a two-inch cable. We
move rapidly, and also our turn time is faster than it is on a
skyline. We move our balloon back and forth at about twenty
miles an hour. Skylines move at approximately that speed, but
take longer to re-rig from one road to the next. Where it takes us
only ten to fifteen minutes, it might take them three to four
hours. That's why balloon logging can be a more economical way
to go.

"Skyline is a good system as far as it goes, but it requires a concave slope. They've got to get their tailholds and spars up high
enough to where they can get a good belly in their line. But most
of those areas, the good concave slopes, are down the drain.
They're gone. We're getting now into the tougher areas where we
have either a fairly continuous slope or a convex slope. In these
tougher areas, they can't get the logs high enough in the air for a
skyline system to work.

"Balloon logging is a different way of life from any other type
of logging. The personnel need to have an interest in it. If you
make mistakes in balloon logging, you're in trouble. You need to
learn it, and learn it well. It's upside-down logging. That balloon
is always wanting to go *up*. In regular logging, you slack everything to get your rigging to the ground to pick up the log. With a
balloon, you *tight-line* it to get it to the ground. If you slack it, it
goes up instead of down.

"In tests we ran for the armed forces, we found that the balloon
that we're using right now doubled its lifting ability in one second
of free flight. When we were just holding it, we were holding
25,000 pounds; when we let it go for one second and then stopped

it, the strain on the lines went to 52,000 pounds. And it went
from rest to 1,000 feet per minute in that one second. You let that
balloon get away from you for a second or two, and you've got a
damned giant on your hands trying to control it. It's a regular
monster. So we just don't break lines. We don't pull tailholds.
And if you do pull a tailhold, the operator has to play that balloon just like a fish. If he puts the brakes on solid, he'll break the
line. The balloon's going like hell up into the sky, and he's got to
put the brakes on slowly just to hang on to it. When the balloon is
loaded with logs, it's a nice tame little pussycat. But when it's
free, it's a damned monster. It's just a giant fish in the sky — and
you've got to play it right."

The choice of a logging system should depend not on preconceived notions, but on a specific analysis of the geographical and
biological features of the site. Balloon logging sounds great, and
it certainly has its place. But, if applied under the wrong circumstances, it can also be out of place. Balloon logging can be wasteful when applied to terrain that requires uphill yarding; yet,
downhill yarding can be environmentally disastrous, if the construction of landings and haul roads will push large quantities of
sediment into the streambeds. Balloon logging can extend up to a
mile into the woods, but such a lengthy stretch of cables is generally too cumbersome for selective or shelterwood logging. Balloon
logging seems ideally suited for rugged, sensitive or inaccessible
mountain terrain, but wind and snow, so common in the mountains of the West, are the balloon's worst enemies. The logging
balloon is an ingenious and valuable tool, but like any other tool,
it has both its uses and its misuses.

THE CHOPPER

The ultimate logging machine is an airborne, infinitely mobile
yarder: the helicopter. The lore surrounding the logging helicopter assumes mythic proportions. No matter how tough the
terrain, no matter how remote the area, the chopper can gently
pluck the trees from the woods. Reaching down for a log here and
a log there, the helicopter can practice extremely selective logging and never disturb the ground. Helicopters have captured the

An airborne yarder:
logging by helicopter

imagination of environmentalists, because they do not tamper
with the earth itself. And helicopters have captured the imagination of loggers, because they provide instant access to the inner-
most reaches of the forest.

Is the helicopter an unmitigated boon? Can it solve our environmental problems, while simultaneously filling our sawmills
with logs? Its benefits are obvious, but what are the costs?

A brand new logging helicopter with a hefty payload capacity
will cost somewhere between five and ten million dollars. A half-
dozen logging balloons, a dozen cable systems, or several dozen
logging tractors could be purchased for the same price. And the
capital expense is only the beginning. A logging helicopter consumes about 525 gallons of gasoline per hour, or over 4,000 gallons for an average working day.[4] The crew is twice the size of a
normal cable show, because the helicopter requires not just a
single operator but a pilot, a copilot, alternate pilots, and a team
of mechanics to service the machine while on the job. The high
capital expense, immense appetite for fuel, and costly labor requirements result in operating costs of about $2,000 per *hour*.[5]

Yet for all this money, the helicopter is not always working.
Since it operates in the air, the helicopter is more vulnerable to
the weather than are other logging tools. The logging season in
the Northwest generally extends from April to October, yet for
about 35% of that time helicopters are unable to operate because
of wind, fog, or rain. Even while on the job, the helicopter must
stop for refueling twice each hour and for maintenance inspections several times a day. Not infrequently there are delays in
locating the logs and attaching the payload to the hovering yarder, and sometimes the load must be returned because it weighs
too much or is improperly attached. Altogether, the helicopter is
successfully operating only about 50% of the available time.[6]

Because of the high capital expense and labor costs, helicopter
loggers feel pressure to maximize operating time and minimize
down time. There is a natural tendency to stretch the helicopter
to its limit, a constant push to increase speed and productivity. If
the time required to yard each turn of logs can be decreased by a
mere fifteen seconds, the daily savings to the operator can
amount to about $1,000.

The pressure on the workers is understandably intensified. They are dealing with expensive equipment, and precision and efficiency are at a premium. As in any work situation, there is an unending tension between production and safety. As the workers adjust their pace to the timing of the helicopter, they are tempted to take chances they might otherwise have avoided.

Perhaps the most dangerous of all logging jobs is that of the hooker in a helicopter show. In attaching the long tag lines from the helicopter to the chokers that have been set around the logs, the hooker must play with bouncing cables and dancing logs. He works under the intense rotor-wash winds of the chopper, which blow dust and debris randomly about, and which often detach limbs from nearby trees. He also suffers occasional shocks from the static electricity that builds up on the tag lines. Yet for all his trials, the hooker is the vital link in helicopter logging, the man who must produce with absolute efficiency in order to avoid cost- ly delays. Though helicopter logging is still in its infancy, there are already several cases of hookers being killed or hospitalized. Of course, the pilots are also subjected to great danger, for log- ging helicopters have been known to crash. A handful of pilots have already been killed on the job.

Helicopter logging has environmental consequences that are not often recognized. True, helicopters eliminate the need to build roads into the woods, but they also require landing pads far larger than the landings used in more conventional systems. If suitable sites for landings can be found, environmental damage can be minimal; but the helicopters are often used in rugged ter- rain, and the construction of the landings can sometimes lead to considerable erosion in a concentrated area.

In the long run, one of the biggest problems with helicopter logging stems from the very selectivity of its operations. The heli- copter picks and chooses its targets, yet since its expenses are so high, the timber removed from the forest must be valuable enough to make the operation worthwhile. What this means is that the best trees will be removed, while the inferior trees are allowed to stand. High-grading, one of the cardinal sins of for- estry, is currently enjoying a renaissance, disguised as environ- mentally protective aerial logging. In any form, high-grading degrades the gene pool of a forest, so the net impact of helicopter logging may be the gradual deterioration of the quality of timber in succeeding generations.

Finally, there is the problem of high fuel consumption. Helicopter logging might save one small area of the earth's surface from being disturbed, but only at the expense of environmental degradation in faraway places. The 525 gallons of fossil fuel consumed by the helicopter every hour have to come from *somewhere*. Logging helicopters place an additional demand on existing petroleum supplies, and any additional demand these days can be met only by developing marginal sources of fossil fuel, often at great cost to environmental quality. Indirectly, helicopter logging in the Northwest may cause or contribute to strip-mining, air pollution, and, perhaps, oil spills.

How can we justify this sort of trade-off? How can we weigh erosion in the woods against air pollution 3,000 miles away? How many tons of sediment are equal to the life of one hooker?

Helicopter logging is not "The Answer." It is costly for contractors and dangerous for workers. For environmentalists, it creates at least as many problems as it solves. Indeed, there is some question as to whether it has benefited environmental interests at all. Helicopters have not been used to log the existing commercial forests more gently, but rather to reach into remote, untouched regions — potential wilderness areas, in some cases — and pick out a few choice trees. Under the guise of environmental quality, they have extended the commercial domain.

A MOVEABLE BEAST

Each of the various technologies presented here — cats, horses, cables, balloons, and helicopters — is a method of removing trees from the forest for transport to a sawmill. There is an exciting new alternative to these technologies. About ten years ago, a couple of loggers from Oregon placed a small engine with a circular saw on a set of metal tracks mounted on a log. What the loggers had invented was a truly portable mill. Now, instead of transporting logs to a centralized mill, we can bring the mill right to the logs. And, in many cases, the new portable mills are a lot easier to move around than the logs themselves.

Mobile mills are particularly well adapted to salvage logging. Back when timber was far easier to come by, loggers left significant quantities of perfectly usable wood lying around on the ground. Because timber is more scarce today, the salvage logs are worth more on the present market than they were when the loggers abandoned them years ago, even though some of the wood has rotted in the interim. This demand for timber residue has created a small, decentralized army of salvage loggers who

Taking the mill to the tree: the mobile sawmill

comb the woods for lost logs. But why truck out the entire log when only a fraction of it is still usable? This is where the new mobile mills can be most useful. By turning the normal sequence of logging on its head — by milling the wood before rather than after it leaves the woods — mobile mills have made salvage logging economically viable.

Unfortunately, the mobile mills are poorly adapted to the small, green logs that constitute an increasing percentage of our timber. Mobile mills have to be mounted on each separate log they cut, and it just isn't worth the extra time unless the log is two or three feet thick. True, there are a lot of three-foot-diameter butt logs still lying around, but they won't last forever, and they aren't being replaced. Mobile mills, ironically, are partially dependent on the wasteful logging practices of the past.

Still, mobile mills can be a valuable tool in the future. They are particularly helpful for milling up wood that will be used near the forest. Like horse logging, balloon logging, and all the rest, mobile milling has its appropriate uses — but it is not a panacea. The holistic forester will do well to use the mobile mill — or any of the other logging tools — only when the circumstances call for it. Each decision has to be made as the situation arises, not on the basis of an a priori belief in the beauty or integrity of any single method.

Chapter 6
Silvicultural systems

W HILE countless tools are used in the harvest of trees, only a few silvicultural systems exist to direct that harvest. In fact, only three are in common use today: clearcutting, selective cutting, and shelterwood cutting. In *clearcutting,* all the trees are harvested at once, creating a large patch of open ground in which seedlings can be planted. In *selective cutting,* the trees to be harvested are chosen individually or in small groups; a major portion of the forest is left undisturbed, providing a continuous cover for the land and a source of seeds for the next generation of trees. The *shelterwood method* is a sort of compromise between clearcutting and selective logging: most of the trees are harvested at one time, but enough healthy, mature trees are left standing to provide a natural seed source and a modest amount of shade for the next generation. Once the new seedlings are firmly established and no longer require a protective cover, the sheltering trees can be removed.

The proponents of clearcutting feel that history is on their side: long before modern logging, huge wildfires periodically devastated the forests to create nature's own clearcuts.

The proponents of the selection method feel that history is on their side: long before modern logging, nature practiced selective cutting by culling out the weak, diseased, or aged trees and opening up small patches of light within the otherwise shaded forest.

The proponents of the shelterwood method feel that history is on their side: long before modern logging, small ground fires swept through the forests and removed everything but the hardy, mature trees that towered above the level of the flames. It was nature's way of opening up the brush and preparing the ground to receive the seeds from the tall and majestic survivors.

Which silvicultural system is really the most natural? Which does the least environmental damage? Which is the most efficient? Which is best suited to provide a vigorous start for the up-coming generation of trees?

CLEARCUT CONTROVERSY

Most of the arguments for clearcutting were presented in chapter 2, but I'll summarize them here. Clearcutting offers unparalleled opportunities for centralization and efficiency. All harvesting occurs at a single time and place. Machinery can move around freely without fear of damage to residual trees; once the machinery is brought to the job, it is allowed to produce to its maximum capacity without external constraints. Clearcutting is logging pure and simple. Once the trees are logged, the area becomes more accessible to other management techniques, such as burning, replanting, and spraying. Clearcutting creates a blank slate upon which the modern tree farmer can create a new forest according to his own notions.

Clearcut patchwork
in the Northwest

The most popular argument for clearcutting among commercial foresters stems from a classification of trees according to their tolerance for shade. Young seedlings of tolerant species prosper under the shade of an existing forest canopy, whereas seedlings of intolerant species require direct sunlight to produce satisfactory growth and development. Many commercial species — including Douglas-fir, the nation's most important source of lumber — are classified as intolerant. These trees need the sun, and clearcutting gives them all the sun they could ever want. Clearcutting, say some foresters, is not just desired for its efficiency, it is actually *required* for the satisfactory reproduction of intolerant species.

But does the entire forest have to be cut down for the sun to reach the ground? If only a handful of trees are removed, the sunlight can still infiltrate the vegetative cover. Group selection — where the trees are cut down in small patches throughout the forest — satisfies the biological criteria for intolerant species by allowing the sun to shine on young seedlings. Since alternatives exist, clearcutting cannot be considered a biological necessity.

Nor is clearcutting the functional equivalent of a wildfire. Fire might kill the trees, but it does not immediately remove them. An area burned by wildfire is covered with snags, and with the insect-eating birds that live in them. The ground remains partially shaded from the sun and protected from the elements by these snags and by scorched foliage. The soil in the wake of a wildfire is not compacted or disturbed by heavy equipment. The nutrients lost due to wildfire will soon be replenished by pioneer brush species, the very plants that are seen as enemies by contemporary industrial foresters. Nature's response to wildfire is to cover the ground with vegetation as quickly as possible; industry's response to a clearcut is to keep the ground open and free of "weeds." Wildfire is a part of a successional process; clearcutting is part of an attempt to bypass that process.

Clearcutting might be economically efficient, but the attempts to justify it with ecological arguments fall short. The environmental drawbacks to clearcutting are real, not imagined. When the vegetative cover is entirely removed, ambient forest temperatures become more extreme, hotter in the summer and colder in the winter. The ground is wetter and less stable during the rainy season, but becomes hard and crusty during the dry months of the year. When totally exposed, the soil is subjected to splash and sheet erosion during heavy rainfall. Soil nutrients are depleted when the entire forest biomass is carried off on a logging truck or burned in the aftermath of a clearcut.

Since natural seed sources are usually destroyed, the area must be artificially seeded or planted. Man's chosen seedlings will be forced to cope with the increased environmental extremes and will face heavy competition from pioneer, sun-loving species formerly held in check by the shade of the forest. To overcome the competition, the commercial trees will often require the aid of human benefactors: herbicides will be sprayed over the new plantation, subjecting animal and human life to unknown hazards. In terms of the overall health of the forest environment, clearcutting, except in certain special circumstances, has serious drawbacks.

GETTING SELECTIVE

Is selective cutting more natural than clearcutting? When nature culls out individual trees from the forest, she lets them rot on the ground; she does not build logging roads throughout the woods to remove the fallen carcasses. In selective logging, roads are built anywhere and everywhere. It is difficult (although not impossible) to log selectively with roadless methods, such as skyline or balloon logging. In order to obtain as much timber as could be produced in a 40-acre clearcut, perhaps 100 acres or more would have to be logged selectively. To reach the scattered trees, a maze of haul roads and skid trails would have to be constructed.

Roads, we have seen, are the major cause of landslides and stream sedimentation. How can selective logging be environmentally flawless when it requires more roads than a corresponding clearcut? The area covered by road rights-of-way is forever lost to production, for the roads are used repeatedly to cull the trees. In a selectively managed forest, the timber is harvested as it matures, perhaps every ten or twenty years. The roadbeds themselves never have time to produce a new crop.

In a mature forest, trees tend to protect each other from the sun and the wind. When some but not all of the trees are re-moved, those that are left are exposed and vulnerable. They are more likely to blow over in a storm. Their lower needles are some-times scorched by direct sunlight, and even their trunks can be streaked with sunscald. The repeated entry of heavy equipment into the forest creates serious problems for the trees that are left standing. The residual timber is often de-limbed by its falling neighbors, and the trunks are sometimes scraped by cats or cables. The damaged trees are, as expected, more susceptible to various diseases.

In addition, the frequent use of equipment compacts the soil, causing the residual trees to grow less rapidly. In an experiment conducted on loblolly pines, a tractor pulling a load of logs was driven past some trees six times to simulate a normal logging operation. The experiment took place in wet weather, maximiz-ing the extent of soil compaction. Five years later, the growth rates of the affected trees were measured and compared with the growth rates of a control group. Where the tractor had passed on one side of a tree, there was only a slight decrease in the growth rate. But where the tractor had passed on two sides, the decrease was 13.7%; on three sides it was 36.3%; where the tractor had passed by on all four sides, the trees had lost 43.4% of their ex-pected growth.[1]

The damage to residual trees is subject to some control. If the trees are selected well and logged carefully, negative effects can be minimized; but there is still the problem of regeneration. How can the seedlings of shade-intolerant species be appropriately nurtured if the forest canopy is not removed? If only one or two trees are logged from any given location ("individual selection"), the growth of the seedlings might be relatively slow. If several trees are taken from the same area ("group selection"), the sun will reach the ground and regeneration can proceed apace. But where do we draw the line between group selection and a clear-cut? How many trees have to be taken before the ground begins to suffer from overexposure and other problems created by clear-cutting? As selective cutting moves toward clearcutting by increas-ing the number of adjacent trees that are logged, the problems of one are replaced by the problems of the other.

Historically, the greatest problem associated with selective cutting has been "high-grading," the taking of only the most desirable trees. As long as there was old-growth timber for the asking, loggers saw no reason to take anything but the best. But taking the best means leaving the worst, and that's exactly what they did. Residual trees were left to re-create subsequent generations, but these parent trees had been inadvertently selected by a sort of reverse genetic engineering. Today, we have inherited the results of this high-grade selective logging: throughout the back-country we are left with low-quality residual timber and natural regeneration from inferior parent stock.

Selective logging, of course, does not necessarily entail high-grading. If the harvested trees are chosen correctly, the quality of the future forest can be improved rather than degraded. Indeed, a well-managed selective system has many positive attributes. When seedlings are nurtured in the partial shade of a protective forest canopy, they tend to grow upward instead of outward. They do not put their energy into the sizable limbs that characterize trees which grow in the open, and their wood is consequently less knotty. The same shade that tempers their growth increases the quality of the lumber which they will eventually produce. The grain is relatively straight and fine, and the wood fiber is actually stronger. Commercially, this means that a selectively managed stand of timber can actually yield greater financial returns than a comparable stand grown in the wake of a clearcut.

There tends to be a greater variety — and therefore greater security — in a selectively managed forest. The trees are of different ages, at different stages in their growth cycles. They root in different soil strata and utilize different proportions of nutrients. Instead of depleting the soil of particular elements heavily required at specific stages, they tend to utilize the soil in an even, well-balanced manner. The accompanying vegetation is generally more diverse in an all-age stand: a shade-loving understory; tolerant species of trees amidst the crop trees; and, a sprinkling of pioneer brush species in the partial clearings.

With greater natural balance, the ecosystem requires less tinkering to keep it in line. Since the danger of insect and disease epidemics is lessened, fewer insecticides and pesticides will have to be used. Since there is no point at which brush threatens to take over the entire forest, there is less dependency on herbicides. Since seeding will often take place naturally from local, well-adapted parents, there is less of a need for elaborate nursery and seed orchard programs.

Finally, selective management systems tend to protect the soil from the ravages of overexposure. There is less nutrient loss due to leaching, less sheet erosion, and less runoff during storms. Since a partial cover of trees is left after each logging operation, there are always roots to hold the soil intact, and leaves or needles to transpire excess water; the ground is, therefore, less likely to collapse from saturation. Although the soil as a whole is not as wet, the upper layer of earth can actually retain moisture better after a selective cut than after a clearcut. Since there is still a vegetative cover, and since a layer of litter still protects the earth, there is less surface evaporation from direct exposure to the wind and the sun. During the dry summer months, the ground tends to be damper and cooler under a forest canopy than out in the open. Seedlings are less likely to die of thirst, and a fire is less likely to break out. And, since moisture and temperature levels are more constant in a selectively managed forest, the soil environment remains hospitable to friendly fungi, bacteria, worms, and other organisms that decompose litter into humus. The normal life of the earth's surface, in other words, can be maintained better by selective cutting than by clearcutting.

Bottom-line loggers are not likely to opt for the selection method simply to protect the soil from environmental extremes or to preserve natural balancing mechanisms. Selective logging, they aver, is too expensive. Each tree to be cut has to be singled out in advance. A road system has to be planned that will reach to every piece of timber, yet the roads cannot interfere with the trees that will be left intact. Tree fallers have to take special care not to damage the residual trees during their work. This requires skilled labor and extra time in preparation. Often, trees have to be rigged with cables or felled with hydraulic jacks to change the direction of fall away from their natural lean. Fallers, buckers, and choker setters have to cover more ground in their work, as do the cats that remove the logs. Since the machinery is decentralized over a larger area, there is a decrease in operating efficiency. Even the cleanup process can be longer and more costly, because broadcast burning is impossible, and the slash has to be bunched in small piles before being set on fire.

These arguments sound plausible, but other variables tend to make selective cutting cheaper, in effect, than clearcutting. In a clearcut everything must be removed, including brush and undersized trees. This involves extra time, and time means money. Marginal trees are hauled off to the mill, even though there's little or no profit in them. In a selective cut, on the other hand, the noncommercial vegetation requires a minimum of handling, and only the larger trees are taken away. Efficiency increases when all effort is focused on more valuable commodities.

A study of logging costs for second-growth ponderosa pine in California concluded that group selection was actually slightly cheaper than clearcutting, and that single tree selection was only slightly more expensive. The total variation in cost between the cheapest and the most expensive method was only 10%.[2] A federal study prepared by the Forest Service and presented to Congress in 1976 concluded that the selection method was generally more expensive than clearcutting, but the difference was only a dollar or two per thousand board feet (the precise difference varied slightly, depending on the tree species).[3] When the price of stumpage ranges from $200 to $400 per thousand board feet, a difference of a dollar or two in logging costs is not really significant. Management decisions must be made on other bases.

Economically, both clearcutting and selective cutting can be made viable. Most companies prefer clearcutting, but there are also those that have operated profitably under selective systems. Boise Cascade used computer programming on its 196,000 acres in southern Idaho, and the computer decided that the selection method resulted in optimal productivity.[4] In western Oregon, the Woodland Management Company has made an impressive profit on cutover land by applying its own version of a selective system. Down by Santa Cruz, California, the Big Creek Lumber Company has outlasted local competitors by selective cutting in the redwoods.

BUD MC CRARY

LUMBERMAN, BIG CREEK LUMBER COMPANY

"Our family started lumbering commercially in 1946. When my brother, my dad, and I got out of the service, we formed a partnership with my uncle. We didn't have much land; we were buying timber from other people. We built a sawmill and moved around to different places.

"In the early days, we found we were cutting about 65 to 70% of the merchantable volume. Most people wanted to take their land off the tax rolls in case timber became taxed in the area. They would take off 70% of the volume and leave 30% or less, so it would come off the tax rolls. But we never did get into any clearcutting or any heavy cutting. We started out with selective cutting, and we felt that that was the best way to log, so we stayed in that mode.

"We also used smaller equipment than most people. We started out without much capital investment. Using smaller equipment, we found we were doing a lot less soil movement; so we stayed with that through the years. We found out that we could log cheaper with smaller equipment in some cases. Not cheaper in manpower, but cheaper from the standpoint of investment in equipment and fuel. Most people don't look at it that way. In the end, we spend more on labor, but we save money in equipment and overall costs. Today, it does cost us more money to log than people up north pay out, but we achieve better results by far.

"The way we practice good forestry on these steep hills is with labor-intensive forestry. We do things the hard way. We're not trying to modernize as much as we should, I guess, but we seem to be surviving. We feel it's important to give people jobs in this area. There's just a shortage of work for people, and if everybody tries to economize on labor, you run out of jobs for people and you've got a bunch of machines and nobody around to run them.

"You have to have darn good fallers to practice the kind of selective harvesting we do, because you don't want to damage too many of the smaller trees. You can't just send some joker up there and say, 'Here's a chain saw. Go up there and fall that mountainside.' You've got to have somebody who really knows what he's doing. We also do a lot of cable rigging and jacking to pull the trees the way we want them to go. That all takes more time and more labor.

"We lay out our logging roads in such a way that the tractor doesn't always get very close to some of the logs. So we wind up with two or three choker setters behind the tractors. That gets expensive, but we feel it's well worth it. We feel that our careful tractor logging will actually create fewer erosion problems in many cases than cable logging. We think we've proven that, and people are beginning to agree.

"More than the cutting, it's the excavating that is really damaging our resources in the long run. We use small equipment and we use planning. We plan the operation carefully. I have three foresters working for me, and we only log about ten million feet a year. But I've got three foresters out there, both making timber contracts and managing the logging operations.

"In most cases the foresters with other companies lay out the original plans; then a logging superintendent comes in and logs it to get the lowest possible cost. In our case, the forester is responsible from the time the job gets started, through the execution of the job, to the follow-up for the job, which takes us another three years. Three years after we've finished cutting we're still in there maintaining erosion control.

"I've trained my foresters in erosion control. It took years of training, but they're able to do it now. I learned this the hard way, and I had to pass it on to them. You go in and size up an area and decide where you want to put your roads, first of all, and see whether or not you can get a road into the area. Then you start developing a road system that will minimize the amount of excavating you will do. That's the most critical part, and that's where you cut down your costs. If you can engineer your road well enough, you can reduce your excavating, which reduces the environmental impact and also reduces your costs. You minimize the width of the road on steep ground. If you have a D-6 and use it properly, you can do an excellent job.

"The real environmental damage, the damage to the soil, started when the D-8 cats came out around 1935. After World

War II they started building stronger tractors, and we had more damage. Then out came the D-9, and we had even more damage. People just didn't recognize what was happening. I don't know why, because when we first started logging here on the Hoover ranch, we recognized immediately that the little gas-powered tractor we were using — equivalent to a D-4 — made a lot less tracks in the forest than a D-7 or D-8 would make.

"Through the years we have stayed with narrow blades. And we sharpen our blades. We actually try to cut our roads with very, very sharp equipment. It's like a surgeon using a scalpel. If you make sharp and clear incisions, then you make very light incisions, whereas if you use a blunt instrument, it takes a lot more power to operate it, to push it through. The equipment doesn't handle as well, it's off. You don't build a road just exactly where you want it, so you're moving more soil to get the same results. When you use a dull knife, it doesn't go where you want it to go. The same thing with a tractor. If you have sharp blades, real dagger corners at the end of the dozer blades, you can put that thing right where you want it. That first pass goes right exactly there. You don't have to dodge off, go above or below a rock, you go through the damn thing. You'd be amazed at the kind of roads we build.

"We also build our roads so that a logging truck just barely fits on them. We figure there's going to be one truck on there at a time. We put turnouts every so often. You put the turnout where you have a wide spot so it won't damage the ground. We use those cheap Mickey Mouse radios, CB radios, and the truck drivers communicate with one another when they're going in and out of these roads. The guy might have to wait two or three minutes for a truck to come out, but we don't have the big highball operations, where the trucks have to pass at forty miles an hour. If you try to build two lanes of road around a steep, 70% sidehill, you're going to excavate probably eight to ten times as much as you would if you were putting a single lane in, a minimum-width single lane, because the excavation tends to square as you increase the depth and height.

"Another part of our operation is a plan for maintenance. You try to decide where your water is going to go, the water that runs off the road. If you don't really know where that's going to go, then it's hard to build a system that will stay in there. So the forester learns through experience where to put those water bars, whether to dip the road in or tip the road out, so the water doesn't run off the side of the fill. It always runs into solid ground. You try to dump your water into a stump, or group of trees, or a rock, some place that will slow up the water velocity. Once a person learns this, he's got most of his maintenance work pretty well done.

"Erosion breaks will sometimes fail; you get a little landslide or something will fall off the bank. Then we come into the area of winter maintenance. After a major rainstorm, we always send crews out in our new logging operations to make sure the erosion breaks are all in and operating properly. We do that for three years, and by the end of that time everything is pretty well solidified. You could have a problem after that time, but if an erosion facility lasts for three years, a lot of vegetation will be established by that time.

"It takes a lot of time and thought to do these things, and it is costly, but that's part of the cost of doing business, and doing it right. Of course there's no such thing as preventing erosion. Even though you leave it natural, the hillside is still going to get damaged eventually. A large, overmature tree will become uprooted under its own weight, and when it lets go, it takes out a big chunk of sidehill. You can minimize it, you can keep the erosion down to the natural rate, but there's no way to prevent it completely. We have a letter from Marvin Dodge, a Ph.D. who worked for the Division of Forestry, stating that his teams found that our erosion rate on two out of three jobs in the Santa Cruz area is less than natural erosion rates, and one of those jobs was just at what they considered to be natural erosion. Those were jobs that we did ten years ago. It shows that if you log properly, you can minimize erosion.

"When we put in a road system, we don't want it to wash away. We try to make it permanent. We try to build a road system that will act as fire protection for the landowner. We find today that people are hiking through these hills and camping anyplace they want to, and we're constantly getting fires started by campers out in the woods. I can't tell you how many times our logging roads have actually prevented a major forest fire. At least four times in Waddell Creek forest fires have been stopped by immediate access by ourselves or the Division of Forestry. Once you've changed a forest to a second-growth stand of timber, you worry about fire, because you've got a succession of trees coming along and you don't want to lose that chain of succession. It's just like the food chain in the ocean. If you burn out all those six- to twelve-inch trees, you don't have that next generation coming along.

"Those smaller trees are the key to the way we practice logging. Every time we go in for a harvest, we're selectively opening the canopy. We're letting in light for the next generation. We'll go from a thirty-secondth of an inch of annual growth to as much as three-quarters of an inch within two years. That's even with some of the limbs removed, being brushed off when we took the neighboring trees.

"We selectively harvest 60% of the trees over eighteen inches d.b.h., and about 50% of the trees over twelve inches in some cases, depending on whether those twelve- to eighteen-inch trees need thinning. If they don't need thinning, we don't go down into the twelve-inch too much, but we do take them where we feel we need to establish some spacing. Actually, the twelve-inch and some of the smaller trees are the next generation. If you wait a few years, the next time you come around the twelve-inchers will probably be eighteen-inchers. We like to think twenty years is the best rotation. If you could get a sixty-year cycle going, in which you entered three times during that period, you'd have a pretty good thing going.

"There's a 'Head Start' program that's already occurred here, where the old-time loggers came in and clearcut and burned and got us started with second-growth redwood. That's a wonderful heritage, a good start on modern forestry. If you treat that properly, you have a perpetual yield going. We think that if you come in and take out 60% of the trees over eighteen inches now and come back every twenty years, you'll have a continuous supply of eighteen-inch trees. We've got that down pretty good. We establish good spacing in the forest and we have a nice looking stand when we're done. We can fly over a cutover area two or three years after it's done, and it's hard to tell it's been cut.

"Many foresters from northern California say that what we're practicing down here is not good forestry, that it's really aesthetic forestry, not forestry for maximum growth rate. Maybe I'd agree with that, I don't know. We're not sure where our particular brand of forestry is leading us, or how much we are reducing the productive capability of the land by operating the way we do. We'll learn that over a period of time. We have a beautiful forest reestablished where the old-timers clearcut and burned, and now we're trying something entirely new: selective harvesting. We're getting good growth on the residual trees that we have, but I think maybe at some point in time we should think about clearing some of these areas and starting fresh. Maybe we should clearcut every couple hundred years and selectively harvest in between. It's a possibility, anyway. We just don't know."

UNCOMMON FORESTRY

Selective logging has received a boost in recent years by the invention of lightweight hydraulic jacks, which can take trees that lean 20 or 30 feet to one side and fall them in the opposite direction. In 1975 Ray Silvey, a lifetime logger who once set a world record in ax-throwing by hitting 25 bull's-eyes in a row, invented a jack that weighed only 39 pounds, but could tilt over a tree weighing 52.5 tons. Using Silvey's "Little Feller," an operator can apply 10,000 pounds per square inch of pressure with only one hand.[5] Any modern tree faller can now be a Paul Bunyan, pushing trees over backwards and landing them precisely where he wants.

The implications for selective logging are profound. Most trees lean slightly downhill; in the old days this meant they would have to be felled downhill. Now, trees can be felled away from stream-beds and directed into areas where they will do the least damage to residual timber. On steep slopes they can be felled uphill, reducing damage to the trunks caused by ground impact. A sizable percentage of timber has traditionally been lost when large trees shatter on rough and hilly terrain, but uphill directional falling can practically eliminate that. In a typical logging show, this means increasing the usable timber by about 10 to 15%, or several thousand board feet per acre.[6] Logging operators receive higher profits, consumers can buy more lumber, and environmental interests are pleased when the damage to streambeds and existing vegetation is minimized by falling the trees in optimal positions.

Another boon to selective loggers has been the experimental system of forest management developed by Richard Smith of Portland, Oregon.[7] Smith calls his system the "Uncommon Forest Management Program," to be contrasted with the "common," even-age management used by most foresters today. Under the common system, suppressed trees are believed to be genetically inferior and are periodically removed to make room for their larger, more vigorous neighbors. Under the uncommon system, the suppressed trees are believed to be capable of further growth and are left intact, while the dominant trees are removed one at a time. The uncommon system is a type of selective forestry, but there is one important catch: the trees are selected not because they have reached an arbitrary, preconceived size, but only because they are stifling their neighbors. A "dominant" tree might be twelve inches thick or thirty inches thick; the important question is whether it is interfering with the growth of the trees around it. If it suppresses other trees, it will be removed.

The uncommon system is based on the ideas of a nineteenth-century Danish forester, C.D.F. Reventlow. In his book *A Treatise on Forestry*, Reventlow argued that loggers should harvest "all those trees which by unfortunate placement prevent their neighboring trees from making rapid growth."[8] Using this criterion on 427 acres of second-growth timber, Smith has tripled the volume of wood in twenty years, while simultaneously logging off 3 million board feet of lumber. The growth rate on the tract has averaged 872 board feet per acre per year[9] — and that compares favorably with what the large companies can get on similar sites that are clearcut, scarified, artificially planted, sprayed with herbicides, fertilized, and thinned. But the uncommon system requires a personalized approach to the land, for each separate tree must be periodically evaluated by on-the-spot inspection of its relationship with its neighbors.

A SHADY OPERATION: THE SHELTERWOOD SYSTEM

Another type of "uncommon" forestry that has become increasingly common in recent years is the shelterwood system of harvesting and regenerating timber. Shelterwood cutting has come into its own as a response to repeated failures in regenerating timber under the clearcut method. Clearcutting is poorly adapted for tree species that like the shade, but it was supposed to work well for the shade-intolerant species like Douglas-fir. In many cases it did, but in other instances even the sun-loving Douglas-fir seedlings mysteriously withered away on sites that had been clearcut.

Poor regeneration has been a particularly severe problem in the southern portion of the Douglas-fir region and on slopes with southern and western exposures. On a problematic clearcut site near Ashland, Oregon, some Douglas-fir seedlings were planted in the direct sunshine, while others were planted in the shade of low brush, rocks, or dead logs. At the end of two years, only 10% of the seedlings planted in the open were still alive, while over 50% of those in shade had survived.[10] Total exposure was too great; moisture and temperature extremes had taken their toll. Similar results on other clearcut sites have forced some commercial foresters to change their minds; Douglas-fir may not tolerate shade, but it doesn't seem to like too much sun, either.

The shelterwood system is designed to produce partial but not total shade, to afford limited protection for sensitive seedlings. In shelterwood cutting the mature trees are removed in stages, and some semblance of a forest canopy is maintained until regeneration is well established. Typically, the forest is first prepared by thinning out defective trees and some of the noncommercial brush. This creates a mild earth disturbance, which allows seeds to penetrate to mineral soil and opens the way for full-scale operation. Loggers can now enter the woods for the seed cut, removing a major portion of the mature timber, sometimes more than half, sometimes slightly less. Ideally, the seed cut will take place just as the cones are opening during a good year for seed production. The seeds will be scattered about and deposited in the ground as the logs and brush are dragged around the woods. After the loggers have left, the regenerative process will occur under ideal conditions: the forest floor will be partially opened, but also partially protected. There will be enough sun and moisture for the seedlings to grow vigorously, but not so much as to produce detrimental extremes in the microclimate. After five, ten, or perhaps fifteen years, the seedlings will begin to offer protection for each other; they will act as a windbreak, produce their own shade, and transpire excess moisture from the earth. At this point, the assistance of the protective overstory is no longer required, and the "shelterwood" trees can finally be harvested. The forest will now consist entirely of young trees, just as it would after a clearcut. The shelterwood system is therefore a type of even-age management, even though there is a short

Shelterwood cut: a Douglas-fir overstory
plays nursemaid to young western hemlock

period during which there are trees of two distinct age classes. Shelterwood cutting resembles selective cutting in its mainte- nance of a continuous forest canopy, but it resembles clearcutting in its simplification of the ecosystem into an even-age tree farm.

There are several distinct advantages to the shelterwood sys- tem. When done properly, shelterwood gives natural regenera- tion an optimal chance of success. The parents of the new crop will be hardy, merchantable trees that have already passed the test of time by adapting to local conditions. The seeds will find a partially disturbed soil in which to imbed themselves, yet this soil will not be subjected to the ravages of overexposure. As seedlings grow, they will be treated to just the right quantity of sunlight. Tolerant species can be given more protection than intolerant species, for the forest manager can adjust the amount of shade simply by changing the intensity of the seed cut. Since there is always *some* shade during the transition from one generation to the next, the seedlings are not likely to be overrun by pioneer, sun-loving species of brush. And, since there is less competition from brush, there is less need for the application of herbicides to counteract that competition.

Yet shelterwood cutting, like the other silvicultural systems, has its vices as well as its virtues. Because it is an even-age system generally applied to a single tree species, the shelterwood system lacks a natural stability based on diversity. Like clearcutting, it is more likely to require artificial props to keep the system in order. Like the selection system, it is accompanied by the problems that stem from repeated logging over the same areas. It necessitates the building of a vast network of logging roads, and in rugged terrain these roads threaten the stability of the earth. The fre- quent operation of equipment compacts the soil, increasing the damage from runoff during storms and slowing down the growth of trees. Residual trees left after the seed cut are subject to wind- throw and sunscald. Trees that are damaged by equipment or by falling timber become more prone to disease. And when the last of the sheltering trees are removed to open up the forest for the upcoming generation, many of the seedlings that have been so carefully nourished and protected are inadvertently but inevi- tably destroyed.

SITE-SPECIFICS: CASE STUDIES

If each silvicultural system has both virtues and vices, how do forest managers decide which one to use? Sometimes the decision is made according to preconception or prejudice. A Douglas-fir forest in southwest Oregon, for instance, might be clearcut because all the textbooks say Douglas-fir requires sunlight for regeneration, and because it is company policy to clearcut for reasons of operating efficiency. Perhaps a steep slope in northern California, owned by an investor in San Francisco, will be selectively cut because the owner has been told by environmentally conscious friends that clearcutting is the eighth deadly sin. Basing decisions on such preconceived notions can have disastrous consequences. In the first case, the Douglas-fir seedlings may perish from overexposure. In the second case, the slopes may crumble under the burden of a massive road network constructed to remove a handful of trees from the middle of the forest. In both cases, decisions should be based on more specific information and more profound criteria. The specific sites should be evaluated first; only later should choices be made. Of course, there is always some evaluation of a site before it is logged, but often the evaluation is far from objective. It is made simply to satisfy legal requirements for decisions already made in faraway offices.

How can on-site evaluations determine the choice between clearcutting, selective cutting, or shelterwood cutting? Consider the following examples:

(1) In a remote area in the foothills of the Cascades, an epidemic of dwarf mistletoe, a parasitic plant, has broken out. Although the mistletoe is spreading rapidly, it is still confined to the northeastern slope of a ridge. The only "cure" for dwarf mistletoe is to remove the parasite from the treetops, and this is done most easily by harvesting the afflicted trees. Selective harvest would open up the forest canopy and damage some of the residual trees, creating ideal conditions for further spread of the mistletoe. To control the parasite and salvage the damaged timber, the area should be clearcut. Since the cutting will be done on a northeastern slope, regeneration is not likely to be impaired by solar overexposure.

(2) In northwestern California, forty acres of second-growth redwoods are to be logged so a family of urban emigrés will have enough money to build a house on their land. The area is populated with other small landowners, who have become accustomed to noncommercial uses of the forest, such as hiking and fishing. A primitive network of old wagon trails already exists in the forest, dating back to the time when the trees were first harvested around the turn of the century. What sort of silvicultural system should be applied? Selective cutting will preserve the noncommercial uses of the land. It can finance the construction of a house and provide for the possibility of additional income ten or twenty years hence. Construction damage will be minimal, since the old wagon roads can be opened up to service most locations. Regeneration will come easily, despite the existence of a continued forest canopy, for redwood shoots will sprout quickly from the stumps of the harvested trees.

(3) On a southern exposure in southwestern Oregon, a timber company wishes to liquidate its mature Douglas-fir timber and start a new crop of trees. On neighboring sites that have been clearcut, regeneration has suffered from overexposure, and most of the seedlings have perished. In order for these sites to be adequately stocked with Douglas-fir, they have been replanted several times and sprayed repeatedly with herbicides to control the brush. On other neighboring sites, however, single-tree selection has also failed to produce an adequate stocking of young fir. There, shade-loving species of lower commercial value have overtopped the Douglas-fir seedlings. Since regeneration has failed from both too much and too little sunshine, some balance between sunlight and shade is clearly in order. Group selection would provide both sun and shade, but the distribution of light would be uneven. A shelterwood system, however, will provide an even balance of sun and shade throughout the forest floor, offering an optimal environment for seedling survival. If the land is not too rugged and can sustain the construction of roads and the repeated entry of the loggers, a shelterwood harvest can simultaneously remove the mature timber and establish well-adapted seedlings, while subjecting the earth itself to only minimal abuse.

Most management decisions, of course, are not so "clear-cut."

Often, some aspects of a given site suggest one silvicultural system, while other aspects of the same site favor an alternate system. The variables must then be weighed against each other, trading off a negative effect here for a positive result there. And the choice of silvicultural system is only the first among many decisions that will determine the fate of the forest. Will a road be placed here or there? Will the logging slash be piled, burned, or shredded? Will the brush be treated with herbicides, hand-cleared, or left as it is? Will the land be treated with chemical or biological fertilizers, or with no fertilizers at all?

Holistic forestry does have its preferences. It favors the harvest technology that least damages the soil. It tends to favor the silvicultural system that provides the least shock to the forest ecosystem. The holistic forester, however, has no pat answers, no absolute solutions. Each site must be evaluated separately, each problem is unique. The ground rules are simple, elemental. The health of the land is always paramount. The needs of the future must be weighed against the needs of the present. And, environmental engineering should be held to a minimum. Given the choice between two management decisions to accomplish a desired goal, the option that requires the least tinkering with natural processes is preferable.

Part III

Mind in the forest: politics and possibilities

Chapter 7

The politics of timber

HOLISTIC forestry, as it turns out, is no big deal. It is little more than thoughtful, sensitive management of the land. It treats nature as an ally, not an adversary. It considers each site according to specific needs. It is, quite simply, forestry that cares about the future. Why, then, is holistic forestry so rarely practiced? The answer requires a look at some basic political and economic realities in our society.

There are three basic types of forest ownership in this country: public, private, and industrial. Each type has its own set of blinders, infrastructural forces that encourage short-sighted, exploitative practices, while discouraging far-sighted forestry. What are these forces? How do they operate in everyday affairs?

PRIVATE LAND

Nonindustrial private owners control 60% of the forested land in the United States, but much of this is in scattered woodlots with minimal commercial value. In terms of standing sawtimber, these private owners account for only 30% of the national inventory.[1]

On the surface, it would seem that small private owners have a unique interest in the practice of excellent forestry. As landowners, they alone will reap the financial rewards if their forests are productive. They can also enjoy the noncommercial aspects of the woods to the extent that environmental integrity is preserved. Their parcels are generally small enough to permit an intimate, personal knowledge of the local idiosyncrasies of their own particular forests. Armed with first-hand knowledge and experience, small owners are theoretically in a position to make decisions and take actions to enhance both the commercial and environmental value of their land.

According to the classical economic model, the vested interest that private owners have in the productivity of their land will eventually benefit society as a whole. As lumber becomes scarce, stumpage prices will rise, and landowners will be induced to invest in timber production for the future. Deforested land will be planted with trees, while existing forests will become intensively managed. It's the basic law of supply and demand: scarce resources will increase the demand, but this in turn will provide an incentive for landowners to increase the supply.

In practice, however, this classical ideal is not fully realized. The life cycle of a commercial tree crop is generally about a half-century, or a third of a century at the least. So there's quite a wait before the supply catches up with the demand. The self-equilibrating mechanisms of the marketplace are geared more toward daily fluctuations than long-term investments. Indeed, when investments require thirty to fifty years to turn a profit, the law of supply and demand tends to produce negative rather than positive results. If stumpage prices rise today, the immediate effect is not so much to induce investments for the distant future as to encourage liquidation of present inventories. Trees that are not fully mature will be harvested when the price has been pushed up because of inadequate supplies.

In personal terms, the small landowner is not likely to invest significant financial resources in a project that might never be completed within his own lifetime. An investment in forest productivity will increase land value, but it might be decades before the resultant timber could be sold and a profit realized. In terms of cash position, forest improvement is far from a flexible investment opportunity.

So when the price of timber doubles within a year (as it has been known to do), the first impulse of the small landowner is not to plant another forty acres of seedlings. It is to harvest the timber he already has. Thus, his timber is added to the presently available supply but is effectively removed from tomorrow's market. The timber shortage is thereby aggravated in the long run, and prices will continue to climb. Even as stumpage prices reach astronomical levels, the small private landowner shows little interest in tying up capital in reforestation and timber stand improvement. When investments are made, they are heavily subsidized by the government through the Forestry Incentives Program.

The public sector is helping to underwrite investments in tomorrow's timber only because individual landowners working within a free-enterprise economy have been unwilling to make the investments on their own. A survey conducted by the Forest Service in 1970 revealed that 80% of the private landowners intended to harvest their timber without making investments in interim management, while only 5% showed any interest in improving their stands prior to economic maturity. (The remaining 15% were holding their land for nontimber purposes, such as recreation, speculation, or development.) The study concluded: "Most forest owners do not consider timber-growing investments to be sufficiently profitable to take priority over other investment or consumption opportunities."[2]

THE CORPORATE MIND

Private individuals might be hesitant to make long-term investments, but corporations, by their very nature, should have a compelling interest in adding to their total assets in order to ensure future profits. On land owned by timber companies, investments in reforestation and timber stand improvement are common. Companies are effectively increasing their capital to the extent that they maximize the growth of their trees.

Corporations receive a distinct tax advantage by treating reforestation as a capital investment rather than as an operating expense. When their timber matures several decades hence, they will not have to pay the normal income tax on their revenues; instead, the money they receive from the harvest will be taxed as a capital gain on their original reforestation investment. Since the rate of taxation for capital gains is only 28%, while the rate for corporate income is generally 46%, the timber companies are treated to significant financial savings.

The ostensive reason for the capital gains classification of timber revenue is to encourage reforestation investments. Theoretically, the money the companies save from this tax break will be plowed back into tomorrow's timber. In fact, however, the extra profits are also used to buy new harvesting equipment, invest in foreign timberlands, and increase the dividends paid out to stockholders.

Ironically, the *practical* effect of the capital gains tax is to discourage rather than encourage reforestation. If reforestation were treated as a normal business expense, it would be tax-deductible. For every dollar spent on tree planting or site preparation, a timber company would save forty-six cents in taxes. The government would effectively subsidize almost half the costs. But under the capital gains system, reforestation is treated as a capital asset, and every penny, therefore, must come out of company profits. Since the company cannot write off reforestation as an operating expense, there is less incentive to allocate funds for future forests.[3]

Yet the companies prefer to stick with the capital gains system that saves them up to 18% on their total timber revenues. The taxes saved by writing off reforestation as an operating expense would amount to much less, precisely because so little money is actually spent on reforestation. Surprisingly, even the large corporations that are committed to tree farming generally invest less than 5% of their funds in tomorrow's timber.[4]

Why are corporate reforestation allocations so small? When a corporation chooses to invest money in timber, it effectively chooses *not* to invest that money elsewhere. Money invested in another field will probably earn interest or pay dividends on a regular basis. Investment in timber, on the other hand, will have to wait several decades to return a profit. When the profit is finally realized from the timber, the returns must approximate the profits that could have been made from other forms of investment. The revenue from a single crop of timber must be high enough to justify tying up capital in the trees for so many years. In other words, *part of the cost of a crop of trees is the interest accrued to the initial investment.*

The interest on timber, when computed over the entire time it takes for the trees (and the investment) to mature, is astronomically high. For every dollar initially invested, a crop that takes 40

years to mature will have to return $10 at 6% interest, or $45 at
10% interest. If the crop takes 75 years to mature, each dollar
will have to return $79 at 6% interest, or $1,272 at 10% interest.
The longer the company waits to harvest its crop, the larger the
returns must be to justify the investment. When the cost of interest is taken into account, any investment that extends beyond
one generation of trees becomes totally unfeasible. There is no
genuine "long term" in the practical world of business, nothing
beyond a few decades. If a company had a chance to spend one
dollar per acre to build up the soil in such a manner that would
only affect the trees 200 years hence, it would be economically
foolish to make that investment. Unless each dollar increased the
worth of that future tree crop by tens of thousands of dollars, the
company would just be pouring money down the drain.

Because of interest computations, *economic maturity* of timber
occurs long before *biological maturity* of the trees. Indeed, economic maturity even precedes *productive maturity*. Two basic
criteria are commonly used to determine when a tree should be
harvested. The Forest Service waits until the annual growth of the
tree starts to taper off and fall below the "mean annual increment" (M.A.I.). The average annual growth over the entire life
cycle is computed, and when that average can no longer be at-
tained, the productive maturity of the tree has been reached.

Timber companies, however, do not like to wait for the tree to
produce to its maximum capacity. For fiscal reasons, they prefer
to reap the returns from an early harvest and quickly invest in a
new crop. The large annual increment in wood fiber is offset by
the interest being charged to the original investment. For an
average Douglas-fir site,[5] for instance, the best economic rotation
at 5% interest is to harvest every 36 years, whereas the mean
annual increment is not reached until 64 years.[6] By harvesting
the trees in their prime, the timber company ignores approximately three decades of peak growth, but cuts its rotation time
practically in half. Instead of paying interest on the original
investment between the 36th and 64th year, it realizes a profit on
the first investment and starts fresh on a new one.

Ironically, to maximize profits a timber company has to cut
corners on production. The company will get two harvests instead

of one, but each harvest will contain well under half the volume of usable wood contained in a harvest at the M.A.I. The company benefits because the interest per crop is only carried over three decades instead of six, but the public suffers because less wood is produced in the long run. In this manner, we sacrifice the future to consume in the present. We fashion our houses out of wood from trees that have scarcely reached their adolescence. Adolescent timber, of course, is distinctly inferior in quality. It is filled with knots from branches that have yet to break off, and it

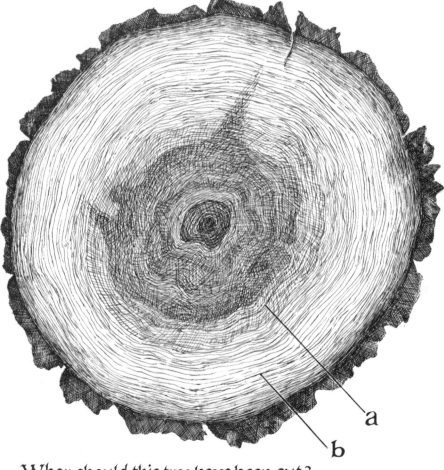

When should this tree have been cut?
(a) economic maturity   (b) productive maturity

contains a disproportionate amount of soft and spongy sapwood, which is weaker and rots more quickly than mature heartwood.

Generally, trees from commercial species such as Douglas-fir must be a foot in diameter before they contain even a modest proportion of quality sawtimber. If Douglas-fir trees are harvested at 36 years of age, however, the yield from 12-inch-wide or larger trees is less than 10,000 board feet per acre.[7] At 64 years, the yield from a similar site would be about 50,000 board feet per acre.[8] Thus, a company can get two harvests within 72 years, but the total output for both harvests is less than 20,000 board feet per acre from quality (12-inch-wide and larger) trees. Strangely, harvesting at *economic* maturity yields less than 40% of the quality timber obtained by harvesting at *productive* maturity.

Why harvest so early? To understand the perspective of the timber companies, let's try to make a typical economic decision. Suppose you have a 40-year-old stand that still has many years of peak growth ahead of it. Should you harvest it now or wait another 10 years? If the prevailing interest rate in the overall economy is 5% (excluding inflation), the trees will have to increase their total output by three-fifths in order to justify the wait. Chances are your trees won't be able to live up to such high expectations. They might increase their volume by one-half, and they certainly will be far more productive than the small seedlings that would take their place, but that is not enough. You, as a corporate executive, are not concerned with production *per se* — you are concerned with money. Since the extra growth can't pay the interest, you cut down the trees.

Let's try this out with real numbers. Suppose, on a perfectly stocked acre of 40-year-old, second-growth redwoods, you have about 58,000 board feet of wood in trees greater than 10 inches in diameter.[9] Suppose, too, that the stumpage price is $200 per thousand board feet. If you sell your timber now, you will realize $11,600 per acre. After paying a capital gains tax of 28%, you will be left with a profit of $8,352 per acre.

What if you decide to wait another 10 years before harvesting your trees? Stumpage prices will probably rise, but let us suppose that they only keep pace with the overall rate of inflation.[10] In real dollars, your timber will still be worth $200 per thousand board feet. The trees will have grown significantly, and that growth is independent of inflation. You can expect a yield of 82,000 board feet per acre on your 50-year-old trees.[11] The sale of timber will bring $16,400 per acre; or $11,808 after taxes.

Is it worth the wait? Probably not. If you harvest your trees at the first opportunity and then invest the $8,352 in another venture that earns a conservative after-tax profit of 5% in real dollars (compounded annually), you will wind up with about $13,600 after 10 years, instead of $11,808. You lose about $1,800 per acre by waiting 10 more years to harvest your trees.

By this same economic reasoning, we can see why the companies must try so hard to minimize their expenditures during the early years of each rotation. Any extra input must produce a much-magnified output, or it simply isn't worth the money. In concrete terms, this means that the companies apply impersonal, area-wide treatments such as herbicides, instead of personal, locally specific treatments such as clearing brush by hand. Area-wide treatments tend to be less expensive, and the savings are themselves magnified by the interest rates. If the application of herbicides is $10 per acre cheaper than hand-clearing, the savings when the trees are finally harvested will amount to several hundred dollars per acre. If the companies don't think that the end product from hand-clearing will be several hundred dollars more valuable than the end product from spraying with herbicides, then they do not feel justified in making that extra $10 investment.

The distant future — and I speak now of centuries, not decades — is also sacrificed by corporate economics. With no economic incentive to plan beyond the next crop or two, investments in soil structure or erosion control cannot be justified financially. Again, the consequence of neglect will be an eventual loss of productivity. Environmentalists claim that the timber companies are acting unethically by ignoring the distant future, but the problem is actually fiscal, not moral. Given the fact that a corporation is an economic entity, why should it invest in activities that show no financial reward? Perhaps it will make token gestures, but these amount to no more than charitable contributions or affirmations of good will. There is no *structural* reason for a corporation to practice the kind of forestry that will lead to a healthy, productive stand of trees 200 years from now.

Nor is there any structural reason for a company to pay heed to the nontimber aspects of forestry. Water quality and wildlife habitat have no place on the balance sheet; in economic terms, they are totally irrelevant. Why should a company modify its timber management practices for the sake of a few fish? The fish might be of economic benefit to others, but unless the company itself can reap a financial reward, it has no business spending good money on the maintenance of spawning beds. Vague and amorphous forest uses, such as aesthetic delight or relief from urban tension, are even less likely to show up on the bottom line.

Financial goals take precedence over the biological needs of the forest. The nation's forests can only produce a finite quantity of timber, yet the companies are constantly pushing to expand the annual harvest, both on their own land and on land belonging to others. Since the industry itself owns only one-seventh of the country's forests, it is forced to rely on outside sources of timber. And since half the standing inventory of timber is found on government land, the companies must lobby for high harvesting rates from the public domain. Pointing to our increasingly serious housing shortage, the industry claims we must turn more trees from the National Forests into lumber. Pointing to high unemployment within the timber-producing regions, the industry claims we must increase our harvests to keep the mills from closing down. Pointing to pressing resource limitations, the industry claims it must charge higher prices for its products.

Whether or not these arguments hold any validity, they constitute after-the-fact rationalizations of positions that are economically dictated. The companies *must* take these stands, just as they *must* maximize their profits: it is the very nature of business to create business for itself. Timber company executives, however, are not so seriously concerned about domestic housing shortages, or they wouldn't ship one-sixth to one-fourth of the timber harvested in the Northwest to Japan every year.[12] They are not that worried about domestic unemployment, or they wouldn't export timber that has not been processed by American labor, nor would they so willingly replace workers with machines. Their public postures are predetermined by their economic roles, not by humanitarian principles or rational derivations. And there is no reason to believe that these economically determined public postures will benefit the forests.

The modern timber company is a vertically organized conglomerate that stretches "from trees to finished homes" (as some of the companies proudly proclaim). These conglomerates own large tracts of timberland and mills that turn the logs into lumber and plywood. They own sources of oil, chemicals, and minerals that are used in the manufacture of wood products. They own distribution outlets to peddle their wares. They also utilize many of the products themselves. Since they possess an abundance of both land and building materials, it is only natural that they focus some of their energies on household construction. They have organized many large-scale residential developments. Some of the companies have entered the growing field of "manufactured housing" in a big way. And with all that interest in homebuilding, the people who started out by harvesting timber out in the woods have quite understandably wound up in the business of carrying mortgages for their residential clients.

The whole economic edifice is entirely rational — but it is based on a logic that has nothing at all to do with silviculture. Financial reasoning leads the companies to harvest timber more frequently than they should, lessening the total production of wood fiber. It leads them to ignore the principles of forest succession that should form the basis of sound forest management. It leads them to harvest timber from areas that are too sensitive to withstand the onslaught of heavy equipment, too steep to avoid subsequent erosion, or too exposed to generate a new crop of trees. It leads them to pay little heed to nontimber values, such as water quality, fisheries, and wildlife habitat. And it leads them to skimp on investments that would benefit tomorrow's timber, since the nature of interest rates renders long-term, slow-return expenditures fiscally unwise.

PUBLIC DOMAIN: THE FOREST SERVICE

Even if private and industrial timber owners are unlikely to pursue farsighted forestry practices, the government, as a timber owner, should be able to do better. It should be immune to the marketplace forces that operate on the private sector, free of the blinders of the profit motive. Theoretically, the government is in a position to practice exemplary forestry.

Government controls 27% of the forested land in the United States (21% is in the hands of the federal government, while 6% is owned by state and local governments) and over 60% of the softwood sawtimber.[13] In a nation dedicated to the principles of private enterprise, such a large commitment to public ownership is an anomaly.

Why has the government entered the business of forest management on such a grand scale? Initially, the forest reserves were set aside as a national insurance against the depletion of timber resources. While private and corporate owners were exhausting their own supplies, the government would hold on to significant quantities of timber that could be harvested at some point in the indefinite future. Today, public ownership is seen in more sophisticated terms. The contemporary principles of public management are set down in the Multiple Use-Sustained Yield Act of 1960: the forests are to be managed for a variety of uses, including "outdoor recreation, range, timber, watershed, and wildlife and fish." The Forest Service is supposed to make "the most judicious use of the land for some or all of these resources, establishing the combination [of uses] that will best meet the needs of the American people . . . and not necessarily the combination of uses that will give the greatest dollar return or the greatest unit output." The renewable resources of the forest are to be harvested, but only at a rate that can be maintained "in perpetuity." And the harvesting activities must be conducted "without impairment of the productivity of the land."[14]

Clearly, the government is better suited than private enterprise to meet this set of goals. A timber company has no reason to manage its land as wildlife habitat. A timber company would be foolish to ignore "the greatest dollar return" as a management criterion. A timber company might have an interest in the next crop or two, but "in perpetuity" has no economic translation. The general public, on the other hand, *does* have an interest in nontimber aspects of the forest, even if that interest cannot be financially expressed. Insofar as that amorphous body called "the public" cares about its own destiny, the government has both the right and the responsibility to manage its land in such a way that our distant offspring can command a resource base undepleted by selfish exploitation for short-term profits.

The stated objectives of public ownership are commendable, but are these goals actually met? Do the everyday mechanisms of the Forest Service lead toward the precepts of well-balanced forestry, as set forth in the Multiple Use-Sustained Yield Act of 1960? Is the business of the Forest Service conducted in such a way as to render these goals realistically attainable?

A federal agency charged with the management of over ninety million acres of land, the Forest Service functions in an unavoidably bureaucratic manner. Decisions that affect the vast stands of timber in the Pacific Northwest are made by administrators and politicians in Washington, D.C., who cannot possibly have seen most of the forests they govern. Regional and district managers must depend on appropriations that are politically determined. And political criteria, like economic criteria, are not necessarily coincidental with the silvicultural criteria that ought to form the basis of sound forest management.

Even when politicians and administrative officials genuinely care about the fate of the forests, the translation of sound silvicultural principles into viable bureaucratic codes can be a herculean task. Administrators with the best of intentions have found it difficult to ensure good forestry simply by changing the letter of the law. They are required by the very nature of their jobs to apply consistent standards to all corners of their jurisdiction. But how can the infinite variability of the forest be reduced to legal terms?

The Forest Service, as a public agency, must establish routine operating principles with respect to all major policy decisions. Administrators must decide, for example, on nationwide criteria to determine the length of timber rotation, how often the trees should be chopped down. To attain some semblance of consistency, the Forest Service has established a basic formula that can theoretically be applied to each and every instance: timber should be harvested just as the "mean annual increment" is maximized. Before that point in time, the trees are still growing vigorously; afterward, a new crop would grow better than the old. Harvesting at the peak of the M.A.I. ensures that the land is producing timber to its maximum capacity.

It sounds good on paper, but in real-life situations there may be extenuating circumstances: abnormal market conditions, insect infestations in neighboring stands, the desire to convert timberland to other uses, and so on. So the law conveniently allows that exceptions be made, that the M.A.I. standard need not be followed in all cases. As any good bureaucrat knows, consistency cannot be purchased at the cost of reasonable flexibility.

Indeed, there is a great deal of flexibility built right into the M.A.I. standard. If the M.A.I. for ponderosa pine is calculated using the total cubic feet of wood fiber as the measure, the rotation cycle is around 40 years; if the M.A.I. is calculated only according to the sawtimber that can be obtained from logs over 7 inches in diameter, the rotation cycle is 90 years; if the M.A.I. is calculated according to the sawtimber available from 12-inch logs or greater, the rotation cycle is well over 100 years. The higher the quality of wood that the trees are expected to produce, the longer it takes to maximize the M.A.I. Indeed, if ponderosa pine is grown for lumber clear of knots and sapwood, the M.A.I. does not culminate until the trees are about 200 years old.[15]

Despite the specifications set up by the Forest Service, the rotation cycles can be determined by criteria that have little to do with silviculture. National harvesting quotas are determined by the state of the economy, the rate of unemployment, the demands of household construction, the optimal number of "housing starts," and other political and economic variables. Once the quotas are established, the administrators of the various National Forests must look at the woods to determine how the quotas are to be filled. Technically, they try to live up to the M.A.I. standard; in practice, they are first told how much they should harvest, then they turn to their inventory statistics to rationalize the allowable cut in silvicultural terms. Perhaps it is justifiable to determine the allowable cut according to the optimal demand for wood products that the government hopes to create, but such a decision is not based on the needs of the forest itself. Because the

Forest Service is used as an instrument of public policy, the application of strict silvicultural standards must often be modified.

Try as they might, Forest Service administrators have been
unable to establish operating procedures to remove the practice
of silviculture from the political arena. Back in 1930, the Forest
Service decided it should not have to beg politicians for reforestation funds every time an area was logged. A sympathetic Congress
passed the Knutsen-Vandenberg Act, which set up special reforestation funds with a percentage of the money received from
harvested timber. The Knutsen-Vandenberg Act is still in effect
today; in the wake of every timber sale, money is *automatically*
allocated for replanting. The "K-V" funds, as they are called,
cannot be utilized for other purposes, nor can they be transferred
to other National Forests. Their effect is to increase political independence and local autonomy. The Knutsen-Vandenberg Act
ensures that no area can be logged and then forgotten.

Yet K-V funding, however well intentioned, has not guaranteed an adequate restocking of commercial timber within the
National Forests. K-V money gives no help for areas that have
been deforested by fire, wind, insects, or disease; K-V money does
not provide for additional replantings if the initial planting attempt fails; and K-V money does not stretch as far as it should,
since the amount is based on a fixed percentage of the initial
appraisal — while the actual task of reforestation often occurs
years later at inflated prices. For all these reasons, an area larger
than the state of Massachusetts remains in the "reforestation
backlog" of the National Forests.[16] Allocations to replant these
deforested regions must come from elected politicians who generally have but a limited interest in forestry. Despite K-V monies,
reforestation funding must still compete with other special interests in the political arena.[17]

Silviculture is dependent on politics, and politics is dependent
on people. Despite its attempt to attain bureaucratic consistency,
the Forest Service cannot be expected to implement the ideals of
multiple-use, sustained-yield forestry with perfect regularity. The
forests might be publicly owned, but that is no guarantee that the
people who administer the rules — or the people who work in the

woods — are acting in the public interest. Even on government lands, most of the jobs are performed by private contractors who have no personal stake in the future of the forest. Their jobs might be circumscribed by contractual specifications, but there is always some room for a contractor to bend the rules to suit his personal needs. Since the concerns of the workers and contractors do not necessarily coincide with the concerns of the forest managers who make the rules, some sort of self-serving finagling is bound to occur.

For example, the Forest Service generally harvests timber according to *scaled sales,* in which the contractor must pay a set fee for each log he removes. The contractor expects to make money on the high-quality timber, but he is required to take out the low-quality timber as well. Often, it would be more profitable to take only the valuable logs and leave the logs of questionable marketability on the ground. The Forest Service may stipulate that a contractor must remove — and pay for — any log over eight feet long and six inches thick which consists of at least one-third usable lumber. But what if a half-rotten fourteen-foot log was inadvertently cut in two? Neither half would meet the specifications for removal, and the contractor would not have to waste his time with the low-quality wood. If the contractor doesn't want a log, he can easily figure out a way to cut it up into smaller pieces that he won't be required to take. Technically, logs are not supposed to be cut in half for this purpose; in reality, it happens.

On land that is owned by the timber companies, such problems need not arise. If a company does not have to purchase each log, it will take anything that can pay its way out of the woods. If the log doesn't pay its way out, it will be dealt with somehow. Since the company will need access to its own land for replanting, it is not likely to leave too much of a mess. Since the company has a vested interest in cleaning up after itself, there is no need for elaborate contractual stipulations, or federal inspectors to enforce them. According to a 1968 Forest Service report, the wasted but merchantable wood runs 50% higher on public lands than on private lands. This waste is due to an inherent weakness in the contractual system. "The timber purchaser's aim in removing wood from sales areas," claims the report, "is maximum profit from extraction and processing; he understandably resists doing anything that reduces operating profit, no matter how much it may contribute to the total efficiency of forest management and timber production."[18]

The Bureau of Land Management (BLM) bypasses this particular problem by cruising the timber before the sale and offering it to the highest bidder in one *lump sum*. Since the contractor has already agreed to pay a set amount, he has a vested interest in extracting as much timber as he can, even if it is of only marginal quality. But the logging contractors prefer not to work under this arrangement, for there is no guaranteed profit. If a significant amount of the timber turns out to be defective — and this is often the case, particularly in old-growth stands — it is the contractor who stands to lose. He has already paid for timber that he can never use. Under the log-scale method, on the other hand, the contractor gets exactly what he pays for.

In recent years the Forest Service has attempted a sort of compromise between the log-scale and lump-sum sales methods. The contractor is required to take all logs of a certain size and quality. Each log is scaled, and the contractor pays for what he gets. But instead of paying for marginal logs as they are measured, the contractor pays a nominal per-acre fee, and he is then free to extract all the marginal timber he can find on the land. As in lump-sum sales, the contractor now has a vested interest in maximizing the amount of timber he salvages and minimizing the amount of waste.

This compromise appears workable, but it is still beset by the problems inherent to all contractual arrangements in forestry. The contractors have no vested interest in the overall health of the forest, but they must be made to behave as if they did. The forest managers (e.g., Forest Service administrators) therefore stipulate an elaborate set of specifications that must be adhered to. Timber sale contracts are voluminous documents that prescribe the details for road construction, logging techniques, the removal of debris, and so on. Since the contractors regard such stipulations as "management constraints" that are to be reluctantly obeyed or surreptitiously avoided, an entire bureaucracy of federal employees must be hired to ensure that the letter of the law is enforced. There are log-scalers to measure the quantity of wood that is taken, field inspectors to make sure the work is performed in accordance with specifications, and office personnel to coordinate and oversee the work of the various on-site inspectors. Each federal inspector is a sort of "boss" to the loggers; understandably, the loggers resent having so many overseers. This fosters a certain alienation among the workers of the woods, and it is not uncommon for this alienation to be expressed by willful disobedience or neglect.

Ironically, even the inspectors can feel powerless and alienated. After all, what methods do the inspectors possess to enforce compliance? They can issue orders to correct mistakes, but if these orders are ignored, elaborate bureaucratic procedures must be followed to impose effective sanctions. What type of sanction is truly effective? Ultimately, only two weapons strike fear in the heart of a negligent contractor: (1) a declaration of breach of the current contract and suspension of all activities; or, (2) a declaration of the "nonresponsibility" of the contractor, thereby making him ineligible for bidding on future sales. But to declare a breach of contract hurts the Forest Service almost as much as it hurts the loggers, for the whole timber sale process then has to be repeated. In practice, a breach of contract declaration is about as rare as a two-headed snake; knowing this, a contractor is not likely to be intimidated by its prospect.

As for a declaration of "nonresponsibility," that isn't even within the jurisdiction of the Forest Service anymore. The Small Business Administration can disallow a bid on a timber sale if the contractor cannot offer proof of adequate financial resources, but the Forest Service cannot disallow a bid on the basis of poor performance on previous contracts. In effect, this means that a contractor can act in bad faith on one timber sale without impairing his ability to do business with the Forest Service in the future. The Forest Service is required to sell its timber to the highest bidder, regardless of past performance. Perhaps this is as it should be, for it prohibits favoritism from creeping into the business of selling timber. But the inability of the Forest Service to effectively judge the quality of work performed by its contractors is an indication that it does not possess the power to enforce uniformly high standards in the work that it lets out to private enterprise.

The weaknesses of the contracting system extend beyond the actual logging operations. Governmental managers let contractors perform virtually all of their work: tree planting, herbicide spraying, fertilization, site preparation, precommercial and commercial thinning, and so on. No matter what type of work is being performed, the name of the game is the same: to meet (or pretend to meet) contractual obligations with minimal effort in order to maximize profit. In reforestation work, for instance, the workers are expected to plant about 1,000 trees per day. To plant 1,000 trees that are likely to survive is difficult, if not impossible; so the tree planters are sorely tempted to cut corners whenever they can.

RAYMOND CESALETTI

REFORESTATION WORKER

"I'd like to go back to the beginning and mention how I got into tree planting. I always had an inclination for things agricultural and for the woods. So the first thing I did when I hit the Northwest was pick fruit: apples, pears, and peaches. Then when that season ended, in the fall of '71, I applied to the Forest Service. You know, even back when I was a Boy Scout, I always wanted to be a forest ranger. I didn't get hired.

"Then I went to the employment office and took a job planting trees. I had no idea what it was going to be. I thought we were going to go out to a farm with a wheelbarrow and little trees in burlap bags. I didn't have any raingear or equipment. I showed up at 6:30 one morning and got into a crummy with a bunch of wetbacks. Couldn't see out the windows because of the mud on the outside. We rode for what seemed like hours. Finally we got out and there we were, on the side of a cliff, practically. It was pouring rain.

"We were all greenhorns, right out of the employment office. So they gave us a quick lesson, a six- or seven-minute training session, and sent us down the hill. I didn't know what the hell I was doing. All the trees I planted were bad. I could barely walk around, let alone chop a hole in the ground and plant a tree. Made $8 that day. It was the worst day of my life. The other two guys I was with quit. They just didn't come back the next day.

"But I started watching those Mexicans. They were making about fifty bucks, so I said, 'There's something going on here, and I'm going to find out what it is.' By the end of that week, I had pretty much figured out what was going on. I had no polished technique, but I knew what they were doing to make the money.

"So I stayed with it and eventually learned the tricks. I got taught some of the tricks by the foreman of the crew. Once I got to where I was a good tree planter about two months later, he finally took me aside and said, 'Look, here's how to hide 'em.' I got a very specific lesson in what they call the 'Wino Technique.'

"The main thing you have to learn is how to fool the inspector. Like if you're in nasty ground that's really not suitable for a bare-root seedling, if it's all rotten wood with solid rock underneath it, you're not going to dig a hole and plant a live, two-year-old tree that's going to survive. But you can get them in there and make them look decent. To somebody who walks by, it'll look like a nice, pretty, little tree standing up straight. Of course in two months it'll be dead. Or you can cut the roots so the tree will fit in a smaller hole.

"I play it by ear. Whatever the circumstances dictate, that's what I do. If a job is very high paying and the standards are very high, I will produce the highest quality work you could possibly imagine, a perfect job with no corners cut, no fudging anywhere. But if the standards are high and the pay is low, too low for those quality standards, then I will fudge where I have to to make my money. Let's face it: I have a right to live and to eat. That's when I work for a contractor. If I work for a cooperative, it's different. I'm not going to fudge on working for a cooperative, because I'll only be hurting myself.

"Once I had a job on a contract for Weyerhaeuser. The contractor was getting paid a piece rate, but he was paying us by the hour. So there was no stimulus for the crew to hide trees, and he was wondering why he wasn't making out so well. I was running the crew for the guy, and we wanted higher pay. He said, 'Well, production isn't high enough to pay more money.' I just said, 'Yeah, because we're trying to give you good quality out here. If you want production to go up, quality is going to go down.' But he said if we got the crew average up, then we'd be likely to get more pay. So then I went out and started jacking up the crew average. I grabbed a few of my cohorts and taught them the tricks, and we started padding the averages to try to get the pay up. I taught them how to hide trees the professional way.

"I showed them the slit method, or envelope method. It involves getting in a place where there's low, tight ground cover. Salal will do, but it's not the best. Never where there's bare dirt. You need a tight root system that's in the humus. You stick the hoedad in just below the surface to make a slit parallel to the surface of the ground, about an inch or so down. You pull up a little to open up that slit. You reach into your bag and pull out three to five trees, bend them in half, slide them into the hole, and lightly push the slit closed with your foot.

"If you do it smoothly, it looks just like you're planting another tree. We once had a bet with the foreman, him and the inspector. We were hiding them, they knew it, we knew that they knew, they knew we knew that they knew. I made a bet with them that they wouldn't find a single hidden tree. And they didn't. The two of them were digging like maniacs all morning, and I was laughing my balls off. I must have hid about 150 of them within two hours, me and the guy behind me. They never found one. There was salal and salmonberry flying all over the place. Those guys worked harder than I ever saw them work, just trying to find where we hid the trees.

"The Forest Service mostly pays by the acre, not by the tree. That leads to a special type of cheating. You, as a contractor, are not going to want your crew to plant extra trees. You're not worried about how many trees per acre survive, you're just worried about covering your ground. You want your crew to get the most amount of acres they can in the least amount of time, which means: 'Don't plant any extras.' Of course you can't plant too few, either, or they'll bust you for it. But sometimes a contractor will have a couple of top planters, his hot dogs, who will get up there where the inspectors can't even keep up with them. They'll wipe out a whole bunch of ground real quickly. Nowadays it's more likely that the inspector will get up in there, but it used to be that they didn't. Those hot shots knew how to go out there to the far end of a unit and chop out a bunch of ground. It would get planted, but in a token fashion, a few here, a few there. Dig it up, make marks and tracks so that it looks like you've been around it all, and then get the hell out. Don't even let the rest of the crew get in there, because they'll go in there and start planting trees.

"Now I'm not condoning this practice, and I don't condone stashing trees. I hate it. It stinks. It feels bad. But if I'm out there seventy miles away from my family for the day, working in the stinking-ass rain and wind and everything else, I'm damn sure gonna get my wage out of it, no matter what I have to do to get it. If the foreman had the money in his pocket, I might be tempted to hit him on the head and take my wage home.

"But I do believe in reforestation. The old-timers, when they used to go into Lucky's downtown and say, 'Who wants to plant trees today?' and haul out the guys who were drinking at eight o'clock in the morning, they didn't give a shit. They just wanted the money. I've got a college education. Why go out into the woods and plant trees? It's stoop labor, nasty weather conditions, and everything else. But I've always felt, and I still feel, that it's a worthwhile thing to do. You feel good when you're done with it, even when you've cheated at it. So maybe you've fudged 10% of the day. You cheated. But the rest of your day, 90% of that day, was a noble effort to replant the forest. I mean I can go now and show those places to my kids. The trees are six feet high. I mean, to me that's meaningful. Most of the men feel like they're providing something that's needed — and they are. It's almost an extension of the Boy Scout mentality, which gets a little disgusting sometimes, but that's the way it is.

"But I won't knock the guys who are just into cheating for quick production all the time. Sometimes it's almost impossible to produce what they want, yet you still have to make your money. You know what I mean? You are not out there purely as a charitable gesture. You're trying to make a living at it, see? So when they're asking you to do the impossible, which they often do, what are you gonna do? You do what it takes to make it *look* like you're performing the impossible.

"The most common way of cheating — and it's done all the time — is J-rooting. Simply speaking, a J-root occurs when you open up a hole and shove the tree down in it, so that the roots form a 'J' against the bottom of the hole. There are many variations: true J-roots, curled roots, bunched roots. Anything other than the root hanging in a natural attitude with the dirt fluffed in around it can be called a J-root. If you look at any Boy Scout manual, they show you how to plant a tree. They show 'right' and they show 'wrong.' And a J-root is 'wrong.'

"The Forest Service now, because of all the hassles they've had with contractors in the past, has several pages of instructions to inspectors, defining how the inspector must go about digging the tree to find the J-root. It's so long and involved. A private industry inspector has a lot more autonomy. A government inspector has a manual that he has to follow, and if he doesn't do things exactly by the book, the contractor can call him on it and file a

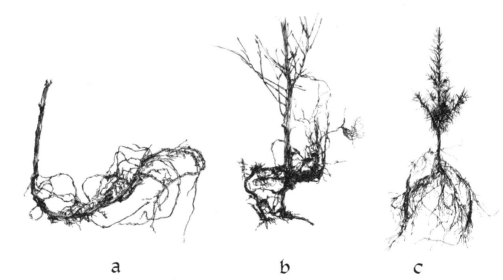

a   b   c

(a) J-rooting, causing slow growth or death
(b) balled rooting, dead within three years of planting
(c) a good root structure

claim with the government. The government specifically describes how many plots the inspector has to take per acre, how many trees in each plot he has to dig.

"And they'll tell him exactly what is involved in digging a tree. He'll get downhill from a plant, go out about a foot, and start excavating. He digs down a good two feet, and then he takes a sharp stick or something and starts very carefully undermining the soil to dig in under the roots of the plant. Hopefully, he'll uncover the roots to the point where he's looking at a cutaway of the planting. But quite often he'll blow it. He digs and digs and his stick catches one of the little roots, and he pulls the whole tree out.

"The Forest Service technique, if they always did it properly, would prevent the inspectors from keeping up with the crew. It's difficult to expose all the roots of a new plant, which has a delicate root system anyway, without disturbing the way those roots are sitting in the ground. What happens is that the inspectors often fudge. If an inspector doesn't find anything wrong, they're going to lean on him. They *know* that people cheat. So *he* just fudges it. He calls most of them good, but he says, 'Well, I found a bad one over here.' The inspector can be lazy just like anybody else."

BUREAUCRATIC SURVIVAL

The government inspectors, like the contractors and workers, have no vested interest in tomorrow's timber. They are performing their jobs according to sets of specifications that have been handed down from above. Like the people whose performance they are employed to judge, the inspectors are ordinary working folk with more of a stake in their paychecks than in the actual survival of the seedlings. The inspectors, too, are subject to the scrutiny of their superiors. Because of the hierarchical structure of the Forest Service, personnel within each stratum must pass judgment on the workers beneath them. With the exception of those on the lowest rung of the ladder, everyone is both a boss and an underling.

How does the bureaucratic organization of the Forest Service affect the work of its own employees? For government foresters,

promotions are slow but steady. Nearly all of the top positions
are filled by professionals who have risen through the ranks. To
obtain promotions, Forest Service employees are subjected to
frequent evaluations of their work. Since promotions can only
be obtained by pleasing one's superiors, the personal ambition
of each worker can become wedded to the policies of the organization. According to an in-depth study of the Forest Service
bureaucracy:

> While the Forest Service can and does tolerate a variety of
> views on particular issues and on particular subjects, it does
> attempt through its hiring, assignments, and promotions to
> develop loyalties to traditional policies of land use and manage-
> ment. While the agency does not consciously attempt to dis-
> courage innovation or new ideas, the incentives created by its
> emphasis on internal promotion and loyalties to institutional
> values favor a fairly conservative and stable policy of land use
> management.[19]

The prospect of professional advancement is the primary
means of encouraging institutional loyalty. It is also important
that promotions be accompanied by frequent transfers. By continually moving its personnel from place to place, the Forest Service ensures that a worker's primary allegiance is to the organization, not to a specific locality. A worker who chooses not to
transfer from his place of work is unlikely to be promoted; consequently, few lay down permanent roots. The managers of our
public forests have no direct, personal stake in the long-term
productivity of the land that is placed in their care. Just as the
contractor may have more of a vested interest in personal profit
than in the long-term results of the work he performs, so may the
forest manager have a greater stake in the advancement of his
professional career than in the fate of our future forests. Certainly, there are many forest rangers who seriously care about the
quality of their work; but the bureaucratic structure does little to
promote a lasting concern for the health of specific localities.

The conservative structure of the Forest Service tends to perpetuate traditional policies and priorities. The people who
advance within the ranks of the Forest Service are those who
identify most closely with these policies and priorities. And the

traditional approach to American forestry places an overwhelming emphasis on timber utilization.

According to a recent survey, 71% of Forest Service professionals believe that utilizing resources is the most important mission of the Forest Service. Only 17% think that the preservation of resources or the maintenance of natural beauty should take precedence over active use.[20] And the primary resource that the foresters wish to utilize is certainly timber.

The foresters' priorities reflect the professional training they received in schools, and the emphasis in American forestry schools is heavily weighted toward the extraction and use of wood products. A study of 37 forestry schools across the nation discovered that all of them required courses in forest economics, forest mensuration, and silviculture, while all but 1 of the schools required the study of wood utilization and technology. In contrast to these timber-related subjects, only 5 of the 37 schools required a course in forest recreation, 5 required watershed management, 4 required conservation courses, and just 1 required its students to study fisheries management.[21] These lopsided curricula reflect the professional interests of the faculty: 4 out of every 5 professors are specialists in either timber management or wood utilization, while 1 out of 5 specializes in nontimber fields, such as wildlife management, conservation, or recreation.[22] It is little wonder that foresters trained in American schools tend "to think and speak of forests as consisting only of trees, and of forestry as essentially the management of forest lands for the production and utilization of timber."[23]

The typical American forester is a man who has received professional training in timber management and utilization, who is paid to use that training, and who is ready and willing to move from one forest to another in order to advance his personal career. Since he generally practices his trade on ten to twenty thousand acres of land, he relies on aerial photography, computers, and calculators to provide him with his data. He has no need to establish a personal rapport with the land he manages. He does not have to consider fish and wildlife habitat, recreation, conservation, or watershed management as seriously as he considers timber. Since he is concerned more with one variable than with the others, he is tempted to reduce the study of the forest to numerical calculations. He is not necessarily expected to see the forest in its entirety and has no personal stake in its future.

THE POLITICS OF REGULATION

Private owners are unlikely to look to the future; industrial owners are motivated by economic incentives that run counter to holistic forestry; the government often seems mired in its own bureaucratic swamp. Is there no hope?

There is an implicit assumption in the body politic that all problems can be solved if only the proper laws are passed. In these days of heightened environmental awareness, this assumption has led to the imposition of increasingly strict regulations to prevent abuses by timber owners and logging operators in both the public and private sectors. Today, there are laws that provide for replanting after every logging job, for watercourse protection, and for a variety of erosion-control measures. Do the laws work? Can regulations and restrictions guarantee good forestry?

On a strictly personal level, the average worker in the woods feels little sympathy for the regulations that are supposed to govern his behavior. Indeed, most loggers are intensely alienated by external restrictions. There is a mystique of freedom about the woods, a mystique shared by the logger, the mountain climber, the fisherman, and all other men and women of the outdoors. To be working in the woods, to *know* the woods, and yet to have one's actions dictated by office-bound persons hundreds of miles away seems the height of folly. The regulations are experienced as personal insults, as a denial of personal integrity.

When a fundamental antagonism exists between the regulators and the regulated, the mechanics of enforcement become rather tricky. Strict and precise standards of behavior must be explicitly stated. Money must be allocated for enforcement officers and on-site inspections. And there must be severe sanctions to effectively prohibit violations of the law. For the regulations to work, all three of these conditions must be met: precise standards, frequent inspections, and prohibitive sanctions. At this writing, there is considerable doubt as to whether any of these conditions are adequately met by current forestry regulations.

California has one of the strictest sets of forest practice laws in the nation. If the laws are to work anywhere, they should work here. In the early 1970s, the California Supreme Court ruled that the existing forest practice laws had to be overhauled, because they were drawn up and enforced by a Board of Forestry that was dominated by the timber industry. The new Board of Forestry had a more balanced composition and proceeded to draft standards that, on the surface, seemed precise and explicit.

The new board defined the different silvicultural systems quantitatively, requiring that a certain number of trees of a given size be left standing for a selective cut, another number for a shelterwood cut, and so on. It required that a specific percentage of the forest canopy be retained within a set distance from streambanks. Utilizing maps of the entire forested region within the state, it labeled each piece of land with a specific slope and soil type and computed a numerical "erosion hazard rating," which could supposedly predict the extent of erosion to be expected in the wake of logging operations. Armed with all this data, the Department of Forestry could presumably require preventative measures to alleviate adverse effects on the environment.

In practice, however, the use of explicit numbers on a piece of paper did not result in the application of precise standards for real-life situations. Despite the quantitative definitions of selective cutting, shelterwood cutting, and so on, many logging operations were approved by state officials under the euphemistic name of "overstory removal," a term that was not written into the law and had no legal definition. Despite the requirement that 50% of the streambank vegetation be saved, inspectors and loggers alike could only guess at the amount of shade preserved, and the requirement only pertained to those streams that had found their way onto official Geodetic Survey maps.

Indeed, the reliance on mapping proved to be a thorn in the side of effective regulation. Streams were not defined by their actual aquatic properties, but by whether surveyors in years past had deemed them significant enough to be officially registered on the maps. Specific slopes and soil types were numerated to describe every square inch of forested land, but in fact the maps only represented approximate averages: the mapping units were composed of ten- to forty-acre minimums, and within each unit the slope and soil types often varied considerably. A forty-acre unit that was registered in the 31 to 50% slope class might have contained ten acres with a 70% slope and another ten acres with only a 20% slope. A fifteen-acre unit mapped as having a stable type of soil might have contained several acres of an unstable soil type, which never appeared on the map.

To some extent, these imperfections in the law can be (and are being) remedied. Imprecision, however, is inherent in the regulatory process. The regulatory bodies cannot be expected to know each and every locality intimately and specifically. They are forced by the very scope of their operations to rely on imprecise and incomplete information. This causes the science of logging regulation to be in fact a pseudoscience.

The Department of Forestry, for instance, may grant permission to log a given area with tractors because it has labeled the area with a low erosion hazard rating. The rating is an actual number, which thus lends a false feeling of objectivity and precision to the entire process. How effective are these numbers in predicting the actual erosion caused by logging? A study of 27 logging operations in California compared predictions from the erosion hazard ratings with the amount of erosion that actually occurred. In only 7 cases did the predictions even approximate the actual extent of erosion. In 4 cases the predictions grossly overestimated the erosion, while in 16 cases the rating system grossly underestimated the total extent of erosion.[24] A rating system as ineffectual as this is hardly a scientific basis for precise and effective regulation.

Recognizing the inherent difficulty in stating standards applicable to each and every circumstance, the writers of the California forest practice rules qualified many of their regulations with "weasel words" (as they are known in the trade): "reasonable," "unreasonable," "minimal damage," "in the best judgment of," and so on. For many of the most important regulations, authority was thereby shifted from the rules themselves to the discretionary powers of the inspectors. Since forestry is as much an art as a science, a good case can be made for the need to bend the laws to reflect the natural variations in real-life experiences; but the substitution of discretionary power for written authority tends to take the teeth out of the laws themselves, for there are always plausible reasons for making exceptions.

In general, California has had trouble codifying and quantifying forest practices in a precise and stringent way. It has also had a problem with the need for continual on-site inspections. The difficulties of inspection in any industry are proportional to the degree of decentralization, and logging is perhaps the most decentralized of all industries. Inspectors cannot satisfy their obligation by an occasional visit to a centralized facility; instead, they must sally forth into the woods to check up on logging shows that are incessantly on the move. The inspectors are supposed to examine each Timber Harvest Plan before it is approved by the state, and they are expected to visit the site again after the operations have been concluded. But to enforce the rules rigorously, they would have to remain on the job when the actual work is being performed, and this would require a veritable army of inspectors. Since the inspectors are government workers, the public is presented with a difficult decision: do we sustain a tax-supported superstructure of nonproductive employees (inspectors), or do we settle for occasional spot checks and incomplete enforcement of the laws?

The final requirement for workable regulations is a set of prohibitive sanctions. Penalties must be established that make it either costly or painful to disobey the law. A gentle slap on the wrist, a mere embarrassment, cannot in itself produce upright corporate citizens; the companies, and the workers, must be made to feel that crime doesn't pay. But is disobedience of a forestry regulation really a "crime"? Can anybody imagine sending a corporate executive or a tree faller to jail for cutting timber too close to a stream? Certainly not. The California law provides for six months in jail or a $500 fine for an infraction of the rules, but since the jail terms are never utilized, nobody takes the threat very seriously. As for the $500, well, a single redwood tree can be worth ten times that much.

The only club with any clout is the provision that offenders will be held responsible for the reparation of all damages caused by a violation of the rules. This could run into money, but is virtually impossible to enforce. How does a logging outfit replace the earth that has been lost to erosion? How can it replace an archeological site that it has inadvertently destroyed? In practice, this provision is enforced merely by requiring the offenders to take more precautionary measures in their future operations. A little good will might be lost by an infraction of the rules, but the show will go on, and business will continue as usual.

With neither precise standards, adequate inspection, nor prohibitive sanctions, regulatory schemes are bound to fall short of their intended goals. This is not to say the forest practice laws have been a complete failure. They have alleviated some of the extreme malpractices that were common in the early days of cat and truck logging: streams are no longer made into skid roads, tractors no longer operate on some of the most unstable slopes, and the forest is no longer lost and neglected after the first harvest.

There are, in fact, several hopeful signs about the politics of regulation. In California, for example, the Board of Forestry has shown that it is committed to the establishment of sound logging practices on a statewide basis. Steps are being taken to remove the "weasel words" from the laws; abuses under the guise of overstory removal are no longer tolerated; the erosion hazard rating system is being overhauled. But is this enough? Will the laws ever carry enough force to counteract the basic political and economic incentives that discourage the practice of holistic forestry?

Under present social forms, there is no concrete reason for a timber manager or a woods worker to pay special attention to the needs of the future forest. Regulatory politics is an artificial attempt to instill such a reason. But it is an inherently negative, alienating attempt. Can we not do better than this? Isn't there some way in which the individual needs of the forest workers and managers can be harmonized with the requirements of tomorrow's timber?

Chapter 8

Landed forestry: a vision for the future

GORDON TOSTEN

RANCHER AND LOGGER

"MY GRANDMOTHER settled in the Ettersburg [California] area in the 1800s. She came out from the East in a wagon train. Ettersburg's where I was born and raised. We purchased the ranch over here [about ten miles east of Ettersburg] in 1929. Timber at that time was valueless, and a lot of the ranchers had decided that ranching was the only thing they could make money on, so they had to destroy some of the timber to make more open land. That's where the girdling came into effect. This ranch had escaped that fate. Some girdling had been done, but it proved to be too much work. Luckily for this ranch, out of the 2,900 acres, it's about half timber and half grass. There was still enough acreage in grassland to make a living on animals, without worrying about girdling trees. Nobody realized at the time that sometime in the future the timber would be worth a lot more than the grazing land.

"Actually, the timber didn't come of any value until the forties. That was when the loggers from Oregon came into this area. They had always said that the timber down here was no good. They said it was too hard, too heavy, a poorer type of Douglas-fir than they were used to cutting. But once the main cut in Oregon had been made, this timber became a little more appealing to the pocketbook. So at that time, in the middle 1940s, the timber on the ranch was sold in an entire block to a lumber company. They started their cut, but they moved out without even removing the timber they had put on the ground. They pulled out to go to a redwood area that was more profitable at that time and they never came back. After five years, my dad just called the contract null and void, and that took care of it.

"My brother and I took over from that point. We were back from the service by then and had just started our own logging operation. Luckily for us we've always had the timber on the ranch, which means we can stay right here at home and run it as a ranch and a timber operation in combination. Because without the timber operation, the ranching just won't make it. There are three partners — my sister, and my brother and I — and without extra capital coming in beyond what the sheep make, well, three families just couldn't live on it. Three families plus my father and mother. We have to log during the good weather, in late spring and summer; and most of the sheep work comes in fall and winter and early spring. So it gives us a year-round occupation.

"We have about 1,600 acres of timberland. It's nearly all Douglas-fir, with a small amount of redwood in one little area. We've never logged any of the redwood, except if a tree blows over we'll go in and pick it up. It's not an overripe type of timber, like some of our Douglas-fir. We've had a lot of old-growth Douglas-fir that was overripe, that needed to be harvested as rapidly as possible — using common sense — to realize the best of the wood before it became more rotten. We haven't logged the young-growth Douglas-fir either, simply because it's growing every year, and that's just like interest on money in the bank. We don't harvest any of that unless it blows down or unless beetles have come in. We've had to remove some of our second growth to counteract the beetle kill. We don't spray. We have no way of fighting beetles other than removing the trees as they die. By removing them as they die, you're still making money. That's actually what most of us are trying to do anyway: make a living.

"We play the market. If this year's market says that they will take a rougher tree, a knottier tree, then that's the type we'll go after. If we have to go for the better tree because the other market is no good, then we'll go for that. You harvest what is needed.

"In years past we used to leave the severely decayed trees. We left those standing because at that time there was no market for them. We knew that someday the paper market would come to this area, and a pulp-type log would be of value. Luckily, we guessed right. Now when we take a log out that's absolutely un-usable for lumber, it is usable — around $35 a thousand — for a pulp log. Every mill these days has its own chippers and its own chip bins to handle this type of log.

"Years ago a lot of loggers made the mistake of harvesting these logs, taking them to the mill, and having them culled. You'd see them piled there on every landing, beautiful logs; they looked beautiful, but they were no good simply because there was too much cull factor. They just went to waste. So one of the wise things we did was leave those trees. We could leave them because we had control of the land and didn't have to go someplace else to wait for another job. We could wait ten or fifteen years for this market, and a lot of people couldn't do that. That's where a lot of ranchers made their big mistake. They would have loggers come in and say: 'We want the entire property logged. We're in the ranching business. We're going to plant it all to grass. We want everything clearcut.'

"Looking back now, we can see that they made an awful mis-take. They know they made a mistake, everybody knows they made a mistake. But it's too late. The damage has already been done. So we have a lot more erosion. They planted their grass all right, but a north hillside will not sustain grass for any length of time, if there was timber on it originally. Consequently, you come back to nothing but brush. Now, of course, a lot of people are coming back to this land, and they've got a tremendous brush problem. They're trying to compete with high tan oak and madrone brush. And this is where the spray comes in that a lot of people are against. Lucky for us, we aren't faced with that prob-lem. We don't have to try to recover brushland to put it back in production, because we didn't create that much brush in the first place.

"A certain amount of brush is needed in order to protect little trees that are growing. So if you have a clearcut area, you allow two or three years to go by and allow the tan oak and the madrone to come, and *then* plant your seedlings so that they get in competition with one another. The seedling will grow faster than the tan oak and the madrone, but those trees will retain moisture in the ground and help give some shade. You don't want any sun-scald. The seedling is growing fast, and his needles are pretty tender. He needs some protection. So you have to have the two together, the seedlings and the brush. Eventually the Douglas-fir will get so high that it will overcome the brush. But a very fine line decides who's going to outdo whom. If you don't plant seedlings quickly enough, the brush will snuff out the Douglas-fir. But if you plant them too quickly, you might lose them to the sun.

Planting by hand: tomorrow's timber

"And animal damage, too. Some of the animals would rather eat a nice, juicy madrone leaf, or even oak shoots, than Douglas-fir. If they have more of a choice, they might leave your seedlings alone. You try to tuck your trees behind some brush where the animals won't get to them. You plant where you think an animal can't get, as long as you see that sunlight can get down to the little tree. We've been planting this way, and it's paying off. We have a lot of them now that are only three or four years old and are up to six and seven feet tall. They were missed by the animals. The ones out in the open that weren't missed are getting bushier.

"Of course, we won't get around to harvesting these trees for many years. We're just now getting into our second-growth timber. Most of the old growth has already been harvested. In the original contract with my dad, it said that we were to remove not less than a million and not more than two million feet a year. The problem is that we cannot have a sustained yield with that amount of cut now that the old growth is gone. We did that for twenty years or better and we still have lots of timber on the ranch, but we see that we can't sustain that kind of a cut. We're now cutting only half a million feet a year. And this year we're putting in a small sawmill, semi-portable, to cut logs that we feel we can do better with than selling to the mill. Whether this will pan out or not, we don't know, but we're going to try it. This will reduce our cut. If we can cut from a quarter of a million to a half a million feet per year and saw that ourselves, I think that we can maintain a sustained yield on 1,600 acres.

"We're cutting less than the original contract called for by quite a bit, but we've got to maintain the timber for our own benefit. We like to see trees stand just as much as our neighbors do. Our new neighbors have told us that they like to see our trees, and they were quite concerned when we told them we were going to cut in certain areas. They said, 'Oh boy, here we go again. There won't be anything left for us to look at.' Then after we cut the trees, we went over to check about some dog problems, and they wanted to know when we were going to be through with our logging. I told them we were through. They said, 'We can't even see where you logged.' So they were pretty happy. And I'm happy, too, because I've got to live on that ranch and I like trees.

My brother feels the same way and so does my dad. It's been that way all of our lives.

"I think the beauty of timber can be maintained, especially if you own your own property. Where you own the property and have control of it, you can make it as bad or as good as you want. Not everybody can be lucky enough to own land for raising both animals and timber. We just happen to be lucky. The timber helps to augment the income from the sheep. If you have a bad year in animals — coyotes or disease or severe bad weather kills off a bunch of your animals — you can say, 'Well I got logging coming this summer. I can maintain what I need.' If you own the property and if you have the right attitude about logging and ranching, you are controlling your own destiny.

"But if you go along with these preservationists who won't let you touch a thing, then you've lost control of destiny. Many laws coming in now make it more difficult and more costly for the local landowner to operate. Not that I'm saying these laws are wrong. We brought them on ourselves. I'm not sticking up for all loggers. There are a lot of bums in the past who really created a big problem. Even us. In the thirty years we've been logging, I can see a great change between the way we were doing it in the earlier times and what we're doing now. We didn't really feel it was too terrible to go through the creek to get the trees along its banks. That was an easy way to do it, and that's the way we did it. Now you find that that's not the best way. But I would hope that the laws don't get too strenuous. I think a lot of the people who are real preservationists are asking for too much.

"We're not a big corporation. On a lot of these shows, the bosses don't work. They sit in the office and they delegate their authority to another fellow who oversees somebody else. We don't do that. We run our own equipment and our own truck. We hire one faller and one fellow to help run the skidder or set chokers. I do all the cat work and my brother drives the truck. We can survive where some people in a bad year would be in real trouble.

"We have a D-6 that we used for the logging. Two years ago we bought a rubber-tired skidder, which operates much cheaper and keeps the logs clean when you bring them in to the landing. We also have a front-end loader to load the truck. And we have a

Kenworth truck to take the logs to market. It's ridiculous when you think that we have somewhere around $300,000 worth of machinery and we're only taking out a half-million feet a year. The bookkeeper just pulls his hair out, because there's no way we can make it. But we make it because we have such a long, protracted contract for the future. We don't have to worry about whether we'll work next year or have to outbid somebody else. All we have to worry about is taking care of our machinery. It goes for years between overhauls. Our breakdowns are very, very few. Because we run our machinery ourselves, we know who broke it down and who to point the finger at. You see, our show is just *different* from the so-called gyppo or the so-called contract logger who has to put out a hundred thousand feet a day, and if he puts out ninety he's gone ten thousand feet in the hole, and that makes so many dollars in the hole. We don't have to worry about that. We take out three to seven loads a day. We take off time to manage our sheep when shearing time comes. We lay down again when we have to ship our sheep.

"We also do all our road maintenance. We have our own grader, a Caterpillar number 12 grader. We've got everything culverted and graveled. We maintain all of our roads on a year-round basis, simply because we need them all winter for checking our sheep. We run no heavy equipment on them in the wintertime. And besides the culverts, we waterbar our roads to try to stop erosion.

"Every fall we go around and open up all the culverts. Then in winter, when the first rains come, we maintain a twenty-four-hour watch, for the first few big storms, especially. That's when all those leaves wash down to the front of the culvert and plug it up. And the wind comes along, and dead twigs fall down in there. If it's a terribly hard storm at night, we go to make sure that none of the culverts is plugged up. If a culvert plugs up at eight or nine o'clock at night, the road will be gone by the next morning. So we go out at night to make absolutely sure they're open, and then we watch them all day, because we're on the ranch looking at sheep. You check them both at once. You look at culverts just like you look at sheep. You maintain both of them. That's why you have so many erosion problems in some of these other logging areas that people love to take pictures of. It's because they weren't maintained. Nobody was out there to ditch with a shovel. After the culverts plug up, the water just goes wherever it wants.

"On some logged land, the owner lives a hundred miles away. Nobody's even there. The fellow that last logged it has nothing to do with it anymore, because he finished his job. He completed his work. He may have put a culvert in, but the owner must keep an eye on it. I don't want to say it's the hundred percent fault of anybody. Sometimes these storms will outguess you, too. You can have a crazy rain that you have little control over. While you're working on one culvert, you can have another one behind you plugging up, and before you get back to it a lot of the damage is done — especially if it's on unstable soil.

"A lot of the problems you see around here, you can't really blame on a particular individual. Just because *we* do it a certain way, that doesn't mean it's right for everybody. I think it's right for us. We're a set of individuals with a piece of land that's ideal. You don't even have to have a yarder on any of it, because it's not too steep. Our creeks are pretty well defined. The way it's laid out, we can log without too much damage to the creeks. And we're a *family* unit. Sure we have disagreement, but we can sit down and hash it over and work it out. That means a lot, too. We're real fortunate. The whole package is just right."

Not every logger, nor every timber owner, can be like the Tostens. Not everyone can expect to inherit a 3,000-acre ranch; and even if one did inherit a ranch, the timber might be gone and the land seriously damaged. The Tosten Ranch is the exception that proves the rule: private ownership can provide the opportunity for a personalized, caring approach to the land; but private ownership also paves the way for rampant exploitation and the immediate liquidation of resources. The Tostens' neighbors were unable to resist the temptation to sell off their timber during the postwar boom. Indeed, so was Henry Tosten, Gordon's father. If the loggers he originally sold to had not greedily abandoned Henry's ranch for better timber elsewhere, the Tostens, like their neighbors, would be sitting on a bunch of cutover brushland.

For those of us who have not inherited a ranch full of timber, the opportunities to break into the business of raising trees are severely limited. Even if we want to practice sustained-yield forestry, where can we find the land to do it? Over half the existing sawtimber in the United States is publicly owned. Since public land is rarely (if ever) placed on the open market, we will have to purchase land from private sources; but the price of country land is greatly inflated these days. The move away from the cities has caused rural land values to skyrocket. The Tostens probably couldn't afford to buy their own ranch at today's market prices (unless, of course, they liquidated their timber resources after purchase). If a newcomer does succeed in breaking into the field of timber management by purchasing land on the open market, his payments will probably be so high that he'll be forced to liquidate much of his growing stock just to meet his obligations.

The most positive aspect of private ownership is that the individual can develop an ongoing, personalized relationship with his land and his trees. In the words of Gordon Tosten, he can "control his own destiny" by managing the land as he sees fit. Since the possibilities for exploitation are always present, however, the owner's control over his own land is circumscribed by myriad regulations and restrictions intended to protect the land from abuse by its master. This conflict — owner control versus environmental restrictions — is a strong undercurrent in timber politics. In the context of American individualism, the restrictions seem like an invasion of privacy; in the context of the damage that can be (and often is) inflicted upon the earth in order to realize a quick profit, the restrictions seem quite necessary.

THE SWISS SYSTEM

If resource exploitation ceased to be financially rewarding, we might be able to transcend the need for strict and severe regulations. Is there some way forest managers could maintain their personal relationship with the land without being tempted to liquidate its resources? Is there some way foresters and forest workers could take a vested interest in practicing environmentally sound forestry without having alienating restrictions shoved down their throats? Is there some ownership/management matrix that encourages personal relationships with the forest, while simultaneously discouraging exploitation for private gain? Yes, there is.

DR. RUDOLF BECKING
PROFESSOR OF FORESTRY AT HUMBOLDT STATE UNIVERSITY

"Historically speaking, the European forests were ravaged in the sixteenth, seventeenth, and eighteenth centuries. Wood was the only fuel source available for growing industry and technology. The forest resource was utilized for everything from steam generation to homes and shelter. As a result, the forests were mismanaged and depleted.

"In about 1720, a group of German professors at a university came together to create a new science, an integrated science, to deal with trees and land, to make land more productive and to grow trees. The foundation was laid at that time for modern forestry. Between about 1730 and 1750, they started planting spruce on deforested lands. An agricultural type of forestry was practiced with these monocultures. These stands grew good and dense, and they were harvested in about eighty or ninety years. That was the first generation.

"After the first generation came along, the biological processes changed a little bit. When they started to plant the second generation of spruce in the mid-eighteenth century, the environment was susceptible to parasites and diseases of the spruce; so the spruce didn't do so well. The balances were upset by this monoculture. Most of these stands became so depleted by parasites and disease that they had to be cut. The explanation was, 'Our technology wasn't advanced enough. If we knew a little bit better about changing the pH of the soil, about selecting better strains of seeds or growing the seedlings better, or putting on some more fertilizer, this wouldn't have happened.' So the stands were planted again with more professionalism and care.

"Actually, the spruce had collapsed because the environment in the lowlands was outside their natural range. There were a few spruces here and there, but it wasn't right for a monoculture of spruce. So by the 1920s, most of those stands had failed again. That can be documented quite well.

"The third generation of spruce was planted right after World War I. The science was then sufficiently perfected, in terms of the technology and the training of professional people. They planted according to strict agricultural techniques. Still, the third generation was not in its natural place, so the spruce failed yet again, this time more quickly. It lasted only about twenty or thirty years. After World War II, that generation of foresters who had tried the hardest came to the conclusion that this was not natural. We had to go back to the natural processes. So European foresters have come full circle in forestry.

"The reason that America doesn't understand this is that most early American foresters were sent to Europe around the 1880s to pick up their forestry. They brought back the idea of even-age management, which we still have in this country. But after World War II, when America became a world power, they did not follow the historical development, the scientific development, in Europe. So at present our forestry does not understand what is happening abroad.

"I received my training in Europe just when some of the professional foresters there were moving away from this idea of even-age management. There's not too much forest in the Netherlands [Dr. Becking's homeland], so 90% of our forestry students had to go abroad to get certificates of professional experience. I chose Switzerland, and that's when I was introduced to the Couvet forest. I became intrigued with the philosophy they had there.

"Their basic theory is that the forest is a community of trees. Experience taught the Swiss many centuries ago that when you start to do heavy cutting, clearcutting, then avalanches will develop, and the whole safety of the valley will be threatened because of these steep mountain slopes. It is of paramount importance to keep trees on the slopes, because that is their protection. They've learned and respected that for a long time. So a primitive management system developed, where they only cut a few trees here or there. Experienced farmers were elected by their community to designate which trees would be cut this year and which trees would be cut next year, and so on. These farmers had no formal university training, but they were in tune with the environment.

"Thanks to two foresters, Adolphe Gurnaud in the nineteenth century and Henri Biolley in the early twentieth century, this primitive system was turned into a science. They carefully measured the rate of growth and the changes in the forest over time. They gave this system its scientific foundations. As a forestry student, I became very interested in the scientific foundations of all-age management.

"In all-age management, you thin young stands out, harvest a few big trees, and create a space for the young stands to grow. You work in all size classes. You cut some noncommercial timber; you cut some commercial timber. The advantage of this is that you can more readily balance your books. You get a repeated income. You get small pocket money every week, not just one lump sum at the end.

"The whole philosophy is different. When you have this small pocket money coming to you, you try to utilize these products better. You don't squander them, because you don't have twenty veneer logs — you only have one. And you're not getting another one until next week. So you are more careful. Artificially created scarcity ensures that you don't waste. You make mine props, pole timber, fence posts, bean stakes, whatever you can get out of it. All the wood gets used, even down to the bakeries that go out and gather the dead wood for fuel.

"The other thing — and this is actually the biggest point — is that under even-age management we have always relied on computerized yield tables. We rely on the table to tell us how the stand should grow, but we have lost touch with the stand itself. In all-age management there is no such thing. You have to establish empirically what changes you can expect from the forest, to see how far you can push the productivity of the forest before it stabilizes on your sustained-yield cut. In effect, you have to make your own yield table for every compartment in the forest. You have to learn the limits of the natural system, and you try to maintain that system. You refrain from capitalizing, amortizing, your growing stock.

"The forester is concerned not only with how he's going to get his timber to the mill, but with what he's going to do to plant it back. The forester is involved in all stages of succession at the same time, as are the worker, the tractor driver, and the logger. Everybody is directly involved. There is no pecking order. There is no status symbol for the logger who takes out the biggest logs.

"In fact, it creates a stable labor force. The same farmers always come back. There are no contracts where they come in to make this amount of money and then disappear. Some of these people always work the same tract. When the forester gets to a certain tract, he hires a certain guy to cut it, because that guy has been doing it since he was a small boy. A tradition has developed. All the way down to the worker, you have a commitment to a specific piece of land. And specific trees. They say, 'Don't touch this tree because I know the old guy who planted this tree.' They have a respect for the old guy, and they translate this respect to the tree, the forest. It's quite a feeling. These guys are very careful. When they log, they know exactly what they're doing.

"The forester has not only to mark the timber, he has to reason out which tree to cut. And he marks which direction it has to go. He has to do it in such a way that his first concern is the replacement, not to lose the growing stock. What can he substitute in its place? He releases some younger trees, which then will grow faster. This is the process — but sometimes he makes mistakes. Then the faller will immediately spot the mistake. He'll recognize which way the tree will go, and, if that differs from what the forester had in mind, the faller can halt the process right away. He can say, 'I don't want to cut the tree this way.' Then there is an investigation made to determine which way it will go. There's a lot more interaction between the boss and the workers.

"The forester is supposed to be able to do it himself if he has to. You can't ask the faller to do it that way if you can't do it yourself. In Europe, tree falling is part of the curriculum for professional foresters. They study the theory, the technique, the safety rules, the forces involved. In Europe, you *compute* the force. You're supposed to be able to compute the force for a tree of this volume when it hits the ground. That way you get an idea of how much breakage you will get. I tried to introduce tree falling into the curriculum here, but I got into all kinds of trouble because of liability. It's too risky. But in Europe you have two years on the apprentice level, and you have the old logger who's going to teach you how to do it. By that time you already have the knowledge of what is technically feasible. You have both book knowledge and field knowledge. And the forester is supposed to know a lot of other things. When the tractor stops, you have to be able to pull the plug and figure out what is wrong. It's all part of the apprentice period.

"In the Couvet forest, the forester and the logger both take pride in how carefully they can bring down a tree. It's built right into the system. You have to save your regeneration or else you lose your prestige, your respect. It's your professional standard. Certain trees that are too big-crowned are earmarked, and then the faller will have to climb the tree. He gets extra money for that. He will saw off the branches and actually lower the branches by rope, so they don't do any damage. He doesn't let

Preparing to fall timber: learning garden forestry
in Switzerland

them fall lest they break some regeneration. Then once the branches are off, he will drop the tree like a pencil into the stand.

"In Europe the hierarchy in the forest service is quite different from what it is in this country. The goal of each forester is to manage his district. That becomes then his home, his source of pride. He uses all his expertise to develop it, to plan the roads, and so on. He is not transferred. He will stay there. He is only transferred when he gets an administrative position — and he usually resists that.

"Foresters are actually kings of their forests. They usually have the right to be buried in their forests. They have the right to put up trespass signs. They are in control of the total forest: the wildlife, the water, the trees. They have the absolute hunting rights. You don't have the right to bear arms in Europe, so you have to have a hunting license. You hunt by invitation. You are invited by the forester to join the hunting party. They hunt on certain days, and they hunt only for certain things. They say, 'O.K., we want to kill three deer: the one with the nick in the ear, the one with this thing, and the one with that.' They try to keep the deer population in balance. They try to figure out how many deer are in the forest. If there is too much pressure on one area, they invite people to hunt those specific deer.

"In Switzerland, the foresters are still elected. Traditionally, if a guy is doing his job he is automatically elected. But he still appears on the ballot. It's a position of public trust. It gives you a status. It gives you power and prestige. And you're supposed to act on the public's behalf.

"The forester in Europe is also the manager of the business. He sets policy and determines what is happening. In this country, the forester is not in charge of the business. That's left to the corporate economist. The forester is like a technician without a professional goal. So the problem is that in many cases there is no leadership. There's no continuity.

"Under all-age management, you have to maintain continuity. When you skid your logs out, you use a constant road. A tree planter has to know that this is the direction the logs get skidded out; he can't go planting his trees in the skid trail. So he has to know his compartment.

"The road network has to be well planned. The roads don't just follow straight lines; they're curved to fit the landscape with the least amount of cut. You have to be gentle with the roads. You don't need a freeway; but you need access to the forest, because the whole area is intensively developed and managed. The roads are small, but they serve a lot of purposes: they are used for fire protection; they provide browse for deer, for wildlife; they let light into the stand for regeneration; they're hiking trails in the summer and cross-country ski trails in the winter; and they provide access to the stand for maintenance or measurement.

"But the equipment is confined to the roads. The main tractor is not allowed to go into the stand *per se*. It can only travel the skid trail. The cable is strung out through the trees to the log, and you winch it in. You stack the logs along the skid trail in piles. The people that walk along the skid trail can see the logs and turn in their bids. You buy maybe five, six piles, or maybe only one. The piles are sufficiently small so that everybody can bid on them.

"The tractor travels the skid road backward, so it doesn't have to turn around and create a tremendous circle. The road is too narrow for that. Also, in Switzerland they do it over snow, so there's no damage to the soil. They wait until the snow hardens a foot or more, and then they skid over the snow on sleds.

"The equipment is generally smaller in Europe than it is in this country. Most of the equipment traditionally developed from the farm. They use a lot of horses in the woods.

"Many foresters have come from the farms. They go off to the university, and then they come back to their homes. They work right along with their fathers. A tradition develops of working on the same land for generation after generation. There is more interest in keeping the land productive."

The Swiss system is, in a word, *personalized*. The foresters and forest workers are dealing with individual trees on limited forest land, not with large, anonymous tracts of raw timber. There is no financial reward for liquidating resources; indeed, the professional standards are based on how well regeneration can be accomplished. The forest is seen as a complete entity that grows

At home and work:
the Swiss forester's stewardship

timber, nourishes wildlife, stabilizes hillsides, provides water, and serves the recreational needs of human beings. The forester is the caretaker — but not the owner — of this entity. He is a "ranger" in the old-fashioned sense: the keeper of the woods. As a public servant and an elected official, he is charged with the task of maintaining a healthy, balanced, and productive forest.

The Swiss system as practiced in Couvet forest is a living example of holistic forestry. The system is not based upon voluminous restrictions imposed by bureaucratic structures. It did not come about by the incessant harangues of environmentalists. Instead, it evolved from a combination of local tradition and a deeply felt need. The need was to maximize sustained-yield production on a small amount of land, to manage the land to its optimum potential, to waste nothing. The tradition was the farmers' bonding to the land over successive generations, and also their respect for the inherent value of trees. Furthermore, the Swiss have historically treated the forest as part of the public domain, the collective heritage of the local inhabitants.

AN AMERICAN WAY

Could the Swiss system be translated into an American form? Tradition, of course, cannot be exported. The forest in this country is not always seen as a collective heritage, nor are commercial trees given the individualized attention they receive in Switzerland. But we are beginning to feel the need. As our own forest resources dwindle, we would do well to develop socioeconomic forms that will encourage the practice of holistic forestry.

MECA WAWONA

FOREST WORKER AND ORGANIZER OF NEW GROWTH,

A FORESTRY COLLECTIVE

"I became involved in forestry mainly through homesteading mistreated land and inheriting the mess, and also through a keen interest in microbiology. I was raising earthworms, and I got the sense that these systems are so delicate and so orderly, and people are screwing around with them on such a big scale out there. It's real dangerous. Once, I took my donkey and my daughter over to the coast. During our forty-mile trek, I hiked through my first clearcut. That did it.

"I started from scratch learning about silviculture and forest politics. I soon realized that silviculture had nothing to do with the way the forests were managed or the decisions were being made. The policies might be based on silviculture, but when it came down to enforcing them and regulating them, and the actual management on the ground, the emphasis was on jobs, cash flow, capital gains, and write-offs. I realized that the only way to really affect forestry was to get involved with labor, with organizing labor and dealing with labor unions.

"During that time I got involved with Redwood Park. The whole issue with Redwood Park was the loss of jobs and lumber. A lot of environmental damage had been noted by the state and the Department of Forestry, but they proposed to repair the area with capital-intensive methods, which wouldn't create any jobs, and all the money would go out of the area. So I worked on putting together a labor-intensive program to repair the lands there, employing some of the people who had been laid off because of the cutback in inventory.

"Working with a grant from the Sierra Club, I produced a report on the labor-intensive approach to watershed repair. Based

on that, we got a contract from the Department of the Interior to
come up with some job figures and some dollar figures. Then I
started meeting with labor union people. At first they scoffed at
my ideas, but soon they started to see the handwriting on the
wall. Now, they're not looking so skeptically at watershed repair
work.

"I tried designing training programs for C.E.T.A. for reforestation work. But they just weren't working. You can't create a
company, you can't create a community, you can't create land-
based foresters through a C.E.T.A. training program. So me and
a couple of my friends decided that we should start our own company to perform this type of work ourselves. We had been working with the silvicultural aspect, but we wanted to get into the
economics. We realized that a project must bear fruit in the end
if it's going to be viable. So we formed a company three years ago
called New Growth Forest Services. Now we have about thirteen
partners.

"Most of the partners live on Greenfield Ranch, which is outside Ukiah [California] in the western hills. It's a 5,000-acre
ranch with about 300 people living on it. The ranch community
provides its own school, its own road system, fire truck, fire fighting equipment, and fire fighters. It does reforestation projects,
trails, and different things like that. And it has communal property that everybody shares: ponds, an orchard, the ranch house.
We have well-drilling rigs and a mobile dimension mill out there.
We've been cutting up an old logjam and building houses on the
ranch with the wood. And the mill will probably build the school,
too. People feel strongly that the ranch is the last frontier for
them, that there's no place to move on to. It's a struggle to learn
how to govern ourselves and tax ourselves for the services we want
to maintain and the land we want to take care of; but that's the
essence of our life out there.

"Our company works on contracts for the Forest Service or the
Park Service; we build wilderness trails and do land rehabilitation work. We try not to get involved in standard tree-planting
contracts, because we don't believe in them. When you plant on
ten-by-ten grids and you eliminate or greatly reduce competition,
you get these fast-growing trees with widely spaced growth rings,

a greater twist, and more knots. Those three characteristics yield a tree that is only useful for peeling for plywood or shredding for pulp and pressboard. If you read the current industry trade journals, you'll see clearly that they're talking about moving toward laminated lumber. Clearcutting and the current type of reforestation that is being practiced are part of a whole scenario aimed at growing plywood and pulp.

"They're turning our National Forests into industrial pulp farms. We've got to deal with this program immediately, because it dictates what kind of products we're going to have in the future. What kind of housing we're going to have, and who is going to produce the materials. Only the largest, most automated, vertically integrated corporations are going to be able to do it. The small mills will go out of business, the few that remain. The conglomerates will be able to control the pricing. And we're going to have to replace our housing every generation. We're going to be doomed to these ticky-tacky, plywood box houses. We are not going to be able to go out into the forests and get logs that we can mill and make our shelter and our furniture out of. There will only be these crappy trees, which we're going to have to let the big guys turn into products that are not going to last very long.

"We'd rather not be a part of that scenario. Instead, we get jobs that we can believe in. When we don't have a contract for trail work or land rehabilitation with the government, we try to get jobs with landowners in the county doing forest improvement work. We've also done work on our own lands through the F.I.P. [Forestry Incentives Program]. We've been training ourselves in silviculture and new types of erosion control techniques. We're doing precommercial thinning and using the poles. We're doing some basic logging work. We're trying to get involved in the full range of activities that we feel are genuinely productive.

"This kind of work is real natural for the people in my area to get into because we're so land-based. We don't want to be commuting to town. A lot of people in the group have carpentry skills and computer skills and different things like that, but people moved up to this area so they wouldn't be commuting back and forth to town and leaving their families at home. People want to be able to maintain their orchards, improve their water systems, build their houses, school their kids, and be able to earn money for the things they can't provide, as well. So they — we — like choosing when we work, in different seasons that aren't so busy for us at our homesteads. Our families are a big part of that scene, too. We pay childcare on our jobs. Our families are in camp. The men and women both work. The kids sometimes work. It's not true child labor, but when we do use them for labor, they get paid. Otherwise they just gather kindling and help in camp with the younger kids and different things like that. A couple of the boys in our company, ages ten and thirteen, are going to be the horsemasters on this next job we're doing. They're going to take care of the packing, because their horses are going.

"So working in the woods blends well with our life-style. And it's meaningful work — improving the land. It's compatible with our politics, our striving toward democratic self-management.

"A vision, a dream of ours, is to develop a system of land-based foresters, a system of stewardship. It takes two directions. One is on public lands. An example is the Ohai system they have in New Zealand. They ask nine adult members to sign a long-term lease — some are thirty-year leases, some are lifetime leases — for stewardship and occupancy of a certain amount of acreage in their national forests. Those people, in return for their use of the forest and their occupancy there, do certain types of management or caretaking activities. The system hasn't been going on for that long, so I don't know how it's working out. But it's something that we've given a great deal of thought to in terms of the National Forests here: a group of people, whether it's a handful of families or a large communal group, having stewardship contracts on a number of acres, large acreage, from hundreds of acres to thousands of acres.

"They could do a number of things there. In terms of maintenance activities, they could do everything from collecting seeds, growing trees, planting trees, thinning, and precommercial logging, to logging and milling. In the more remote rural areas, they could do the milling right there and truck the lumber down to the towns. Where there are large mills in the area, they could either transport the logs out or have trucks come up and take them out.

They could also be doing hardwoods, which require greater understanding of how to cut them, how to dry them, and how to stack them. They could also help maintain fisheries and do monitoring. They could do the maintenance and monitoring for work and research set up by university people. They could also maintain hostels, whereby people could come into the area by trails and participate in a number of work-oriented forest activities, just for the work experience. It would move away from 'Winnebago' recreation, toward learning about the forest and having a real vacation.

"Those communities would be doing their own training of young people in their work, their own schooling, homesteading, food production, shelter production, and those kinds of things, along with the stewardship of the forest resources. I see it as more than an even trade of occupancy for all of those things they're providing. They would get paid for those services.

"Another possibility for land-based forestry would be on private land. You'd have land trusts where the same types of things would be occurring, only in this case a state or a federal subsidy would be required. Some of the subsidy would be paid for by stumpage tax, not just to pay for tree planting, but to pay for all the different types of work that need to be done on land to bring it back to a productive state.

"Within all these land management fields — wildlife, fisheries, forestry, and so on — the technical staff on the ground are noting more and more the need for forest improvement work, fishery improvement work, and all this kind of work, but they don't have the funds to do it. They don't have the staff and they don't have the money to put these projects out. That is going to have to change in the future if we want to have a continuing forest resource. The public must take a new look at the concept of subsidy. Investing in forest resources is not really a subsidy at all. The forest resource is a *natural equity,* a natural reserve bank. When you liquidate your equity, you have less and less interest every year. Those that are liquidating the equity to use it must deposit back into the bank to maintain an equity, so they can keep harvesting the interest. A stumpage tax for forest improvement is simply putting money back in the bank.

"The Greenfield Ranch and the rest of this county was just raped in the fifties in the postwar housing boom. All the ranchers cut their logs. People got cheap lumber to build cheap houses, but that price didn't reflect the real cost of that lumber. The full cost of all of those houses has not been paid. Putting money back into improving those lands and repairing those lands is the cost of the lumber that came out of them. If we keep passing that cost on to our kids, it gets more expensive all the time, in terms of the amount of soil and number of fisheries that we lose, and in terms of inflation. We've got to start paying it now.

"I really believe that the best way to repair the land is with land-based foresters. We learned ourselves, from working on National Forest land and in Redwood Park, that if you live at a site for a while you come to understand its needs, its fluctuations during the day, the way shade moves around, where you plant what trees, all those things that you can't observe as a scientist who comes out of Sacramento or Berkeley to set up a plot and check it once or twice a year. Or as professional foresters who come out in their pickup trucks, cruise the site, and draw up a plan based on that quick cruise. We learned when it started raining how much the creeks rise and which gullies rise and what size the check dams need to be. But you only learn that kind of stuff *if you're there*. You don't learn it if you split the site and go inside as soon as it starts raining.

"My dream is to wake up in the morning and get on a trail and go to work with my family, my daughter, and my friends, not to get in a truck and drive thirty miles like the loggers do now, or like tree planters do now. The way it is now, you commute horrendous hours each day to get out to the forest. It doesn't make sense.

"The ownership lines we have now don't make any sense, either. It's real hard to manage a little plot, like a 160-acre plot, when the lines were drawn in some developer's office with rulers. It had nothing to do with the springs on the land, or the trees, or the hills. To practice good, effective forestry, you need to control a whole forest or an entire watershed.

"To get around the limitations of arbitrary property lines, some of us are starting to talk about developing public resource

corporations. Landowners from an area can set up a public resource corporation with an elected board from the community.
The land is leased — long-term leases — by the corporation. The
landowner gets a little bit of money every year, instead of one
lump sum from one harvest that he or she has to pay a large
amount of taxes on. The land is managed by the public resource
corporation. Seeds are gathered from the right sources. They
have cooperative equipment and a cooperative nursery. Labor
comes from the community; the public resource corporation is
contracted to do different types of work. The landowner has a
more stable income and more stable relationship with his or her
forest resource than under the present system.

"A person's land doesn't have to be logged all at once like it
does now. A rancher who has maybe a thousand acres can't
afford to pay L.P. [Louisiana-Pacific Corp.] to come in a few
times during a decade and get these trees and then those trees.
They have to get it all at once, that's the way L.P. works.

"This would be different. The equipment is right there in the
neighborhood. The crew is working there all the time. Maybe
that landowner is being employed, too. They can afford to do
selective work. They can move in and take maybe twenty trees.
They'll also do thinning. They'll do slash work. They'll do fishery
stocking. They'll do all kinds of things on that site. The moving
and set-up costs are shared by all the different activities and all
the landowners in the area.

"Something like this might start happening within the next
decade, but it's not happening now. There's a constant tension, a
constant struggle in my life, between dreaming for the future and
what's really happening in the present. I can't help seeing the
deficiencies, the real dangers, of present forestry programs, and I
want to go right to the source and change them. But it's hard to
find the time to try to change things. I get paid for swinging a
pick, not for going to Sacramento and telling them how they
could do it better or what the problems and dangers are with the
way they're doing it now. The Sierra Club lobbyists, for the most
part, are from urban areas. They haven't worked with the tools
and the seasons out there in the forest. They do the best they can
in trying to outlobby the industry representatives in terms of for-

est practices, but they can't really get to the nitty-gritty of it and be as effective as forest workers can.

"I see it as the responsibility of forest workers to create change. We've been *given* so much in being able to get back to the land and find that new way of living that's so demanding, but also so enriching. We've been given the incredible richness of our lives on the land, but a responsibility goes with it, and that is communicating what we've learned from the land. That's hard for me, because it takes me away from the land to the hubbub of Sacramento and a not very healthy life-style there. But that's the only way we're going to affect the future."

THE BOTTOM LINE

There is a basic paradox inherent in the private ownership of forest resources: if you own the forest, you possess the means to exploit it; if you don't own the forest, why should you care enough about its future to do it justice? The concept of stewardship is an attempt to bypass this paradox. A steward does not have the ability to exploit resources for private gain, yet there are ways in which a steward can be enticed to act in his own behalf, while simultaneously acting in the best interests of the future forest.

Presently, loggers and timber owners are paid according to how much timber they extract from the forest. What if the forest owners and workers were paid instead according to the growth rates of the standing trees? If the stewards of the land had an immediate economic interest in the timber they were growing for the future, they would naturally take all possible steps to maximize their growing stock. And in the long run, it's the growing stock that counts the most. The determining factor in timber management should not be the quantity of wood harvested in any given year, but rather the quantity of wood that the forest is actually producing.

Imagine three or four families caretaking about two thousand acres of wooded land on a long-term basis, perhaps for a decade or two, perhaps for life. These stewards must harvest a set amount of timber each year, but they do not receive any direct payments for the logs they ship out. Instead, their salaries are based on the timber that is presently growing on their land. The greater their growth rates, the more money they receive.

But why limit the concept of productivity to timber alone? Doesn't the forest also produce fish, wildlife, water, and recreational opportunities? Indeed it does, and the stewards of the land ought therefore to pay some attention to those other attributes as well. Why not base their salaries also on such non-timber variables as the extent of their fisheries and the quality of their water?

Here's how it might work. The stewards are responsible for all aspects of forest maintenance and improvement. They collect seeds from local sources, raise seedlings, plant trees, control brush, and harvest timber. Since they are paid according to an inventory of their standing timber, they naturally try to maximize their growing stock. This means that they will take special care in their logging operations not to damage the land or the residual trees. It means they will try to minimize erosion, for the topsoil is their bread and butter. Every cubic yard of dirt that washes away means less timber — and less money — for the stewards. If the soil settles into the streams, so much the worse. Since the stewards are paid according to the extent of their fisheries and the quality of their water, stream sedimentation is likely to bring them an immediate economic loss. Consequently, stewards must constantly patrol the land to prevent erosion. They must also provide trails, campsites, and perhaps even hostels to accommodate visitors to their domain, for even recreational opportunities should be figured into the complex formula that determines their pay.

The stewards, in short, must keep the forest perfectly primed for all its various uses. To succeed, each must have an intimate knowledge of his own particular place. They must tend the forests with the greatest of care, for they are not mere contract workers or bureaucratic soldiers. The stewards are simultaneously local woods workers and forest managers. They are *landed foresters.*

To be sure, this idea presents its own set of problems. What kind of incentive schedule would include such diverse variables as the quantity of timber and the quality of the water? Who would conduct the measurements necessary to implement such an incentive program? Who would oversee and evaluate the work of the landed foresters?

The first problem is technically difficult, but certainly not insoluble. The quantity of standing timber could be cruised at regular intervals, say every five years. By comparing successive cruises, the change in growing stock can be readily determined. Determining the state of the fisheries is slightly more complicated and less reliable, but the counting of fish is not an impossible task. Similarly, techniques are available whereby the water quality can be measured according to various criteria, such as stream sedimentation or chemical pollutants.

How should we weigh the conclusions of each of these distinct inventories? In order to implement the incentive program, the general health and productivity of the forest must be quantified. Eventually, a landed forester will receive a precise, definite sum that must reflect the work he has accomplished in all fields. Needless to say, this payment can only approximate what is really his due; but there is no reason why the approximation can't be fairly close. And we do not require perfect accuracy to accomplish the basic purpose: motivating the landed foresters to perform good work. Stream sedimentation and incomplete stocking are sure to show up on the balance sheet somewhere, and since a landed forester knows this, he will try to remedy any such problems. Conversely, if a forester tends to the land and manages to increase the quantity of timber, he is bound to be financially rewarded.

What about the variables that are harder to quantify, such as the recreational enjoyment of visitors or a balanced wildlife habitat? Do we need a special team of inspectors to check out the hiking trails or look for signs of distress in the deer population? How can these be evaluated? Wouldn't it take a large and unwieldy bureaucracy to implement this whole evaluation process?

No, it wouldn't. The landed forestry program could evaluate and police itself by combining the inspection process with a unique type of educational institution. Since the landed foresters would be called upon to perform a variety of tasks, they naturally must receive considerable professional training. And who would make the best teachers for this professional training? Naturally, people who are landed foresters themselves. So, let's imagine that a certain number of experienced landed foresters are granted "sabbaticals" from their own land to teach at the appropriate forestry schools. Together, the forestry students and their teachers would visit various places where landed forestry was being practiced for some on-site training. There, the students would learn the techniques of their teachers; the same students would measure and evaluate the work of their hosts.

Under the supervision of their teachers, students would conduct the various inventories required by the incentive program. They could also undertake a basic, impartial evaluation of the nonquantitative aspects of forest maintenance; they could determine, for instance, whether the trails and campsites were adequately maintained.

The central idea here is to transform the evaluation process from a nerve-racking "test" to a medium for positive interaction among all the interested parties. The students would learn from their hosts. The professors would get a chance to examine techniques used by other landed foresters, while the hosts, in turn, would benefit from the ideas and experience of the visiting professors. The three-way exchange of ideas and techniques would help transcend the traditional dichotomy between the judges and the judged. Indeed, these visits might provide welcome intellectual stimulation for folks who spend most of their time working in the woods. Experimental concepts might be explored; new techniques might be developed. This would be more than a routine inspection — it would be a traveling university of the forest. There would be no self-important bureaucrats determining the fates of the workers in the field. All the participants would be practicing landed foresters of the past, present, or future.

Informally, these evaluation/exchange visitations would determine for the landed foresters not only their pay scale, but also their professional reputation. More than money would be at stake in being able to show off a productive, healthy forest. To be regarded as a success in his chosen field, a landed forester might show where he had reclaimed a landslide for timber production or removed a logjam to open a streambed for spawning. He would receive no special credit for routinely cutting down trees; all landed foresters, after all, would have to meet their harvesting quotas. On the other hand, if a landed forester could fall a hundred-foot tree between two young saplings without doing damage to either, that would be cause for admiration.

A landed forestry program could simultaneously maximize timber production and minimize the environmental stress caused by timber management activities. There would be no incentive at all to harvest the timber before it reached its peak productivity. Indeed, a tree cut down would no longer contribute to a landed forester's paycheck, so harvest would proceed reluctantly. Of course, foresters would be required to harvest timber to meet their quotas, but the selection of which trees to cut would be left up to them. They would naturally choose to remove those trees which have already maximized their mean annual increment, or which are inhibiting the growth of their neighbors. There would be no need to draw up rules and regulations to promote productive practices; landed foresters, acting in their own behalf, would automatically try to produce as much of tomorrow's timber as they possibly could.

Accordingly, the landed foresters would utilize all available space. Presently, tiny seedlings are often planted at ten-foot intervals, and the spaces in between lie fallow for at least a decade before they are covered up by the crowns of the crop trees. If anything does grow in these spaces, it is generally removed by herbicides. This in-between land, therefore, remains unproductive for a considerable period of time. But what if some short-lived green-manure trees were grown in these intervals? The landed foresters would have three good reasons for utilizing the extra space: they would promote the growth of their crop trees; they would add these green-manure trees to their growing stock, thereby increasing their paychecks; and they would produce wood fiber that could help meet their harvesting quotas ten or fifteen years hence.

Trees that are presently considered "weeds" could be figured right into the quota system. Since the wood-energy market is likely to mushroom in the near future, there will be a demand for quick-growing fiber producers, such as alder. The harvesting quotas would reflect this demand for wood energy, as well as all other demands. A landed forester would be expected to supply a certain quantity of sawtimber of various sizes, another quantity of peeler logs, a set number of poles, and a predetermined amount of wood fiber. These quotas, of course, would be determined according to an initial inventory of what the land could produce on a sustained-yield basis. The quotas would serve as the vital link between the landed foresters and the marketplace.

The quotas could also serve as an instrument for the development of well-balanced timber stands: to the extent that the quotas require diverse products, the landed foresters would have to maintain some diversity within their forests. This diversity, in turn, would help provide the forest with natural stabilizing mechanisms. To the extent that monoculture is superseded by genuine silviculture, there would be less of a need for artificial controls such as herbicides and pesticides. The brush, for instance, need not be *eradicated,* it could simply be *managed.* With the demand for hardwood built into the quota system, landed foresters would have to cultivate some of these "weeds" along with their conifers.

The landed forestry program would provide an economic rationale for terminating even-age monoculture. Today, there is a strict division between industrial tree farms and preserved park land, between monolithic timber production and the aesthetic or recreational uses of the forests. Under the landed forestry system, a ranger would not have to nag at a logger to preserve the environment, because the ranger and logger would be the same person. Presently, professional foresters make all the crucial decisions and then disappear from the scene while the loggers, tree planters, and other piecemeal contractors are left to do the work. The environmental concerns of the foresters or the regulatory agencies are perceived by the workers as external constraints, rather than positive goals. The landed forestry program would reverse this situation: the very same person, for instance, would prepare the harvesting plans, log the timber, replant the seedlings, and reap the financial rewards for treating the environment with respect. There wouldn't be many catskinners under this system who would wantonly push dirt into the streams, nor would there be many tree planters who would stash trees behind the inspector's back. Destructive shortcuts such as these would no longer be lucrative; indeed, they would become economic liabilities. Only by managing the environment with the greatest of sensitivity could the landed forester of the future achieve financial success and professional status.

The landed forestry system is conceived to protect the environment while simultaneously maximizing the production of timber. And the system has social benefits as well. For the individual workers, landed forestry would mean year-round job security rather than seasonal and/or migratory employment. Their jobs would be varied and challenging, rather than repetitive and boring. Because workers would move from task to task, occupational health hazards would be minimized: tree fallers and brush cutters wouldn't get white thumb from handling a chain saw day after day, nor would planters get tendinitis, nor catskinners develop kidney ailments. Because of the challenge and the variety, the landed foresters would probably not feel alienated by their job situations. And they certainly wouldn't be alienated by oppressive foremen or bosses — they wouldn't have any. The landed foresters would function as autonomous professionals, not as cogs in a bureaucratic wheel. As much as is possible within the laws of nature, they would control their own destinies.

In a sense, the landed forester would resemble the homesteader of yesteryear, in that he would make his living from a particular spot in the forest. To be sure, there are significant differences: the landed forester, unlike the homesteader, would have no interest in liquidating resources, nor would he be able to acquire more land to expand his personal domain. But the landed forester could approximate homesteading self-sufficiency. He could certainly provide his own shelter, and he might well be able to raise much of his own food. He would have a rare and enviable independence. He would be the space-age inheritor of the Jeffersonian tradition. Back in the nineteenth century, the government tried to encourage small, independent farmers and loggers to settle the West. It didn't work. Could it work in the twenty-first century? Could something like the landed foresters program give new life to the Jeffersonian ideal?

THE COMMON GOOD

The landed foresters program would benefit not only the individual workers, but also society as a whole. Presently, the government — and that means all of us — picks up a significant, although well-hidden, portion of the cost of producing timber. Most loggers in the Northwest work during the dry months and collect unemployment when it rains. Most tree planters work when it rains and then collect unemployment during the dry part of the year. Woods work is fragmented and seasonal, and the public is expected to pick up the pieces. Under the landed forestry program, these social welfare costs would be eliminated.

The improved health of the forest can also be considered as a social benefit. We would have more fish, better water, and improved recreational opportunities. We would have more timber, too, and that means cheaper lumber and additional available energy.

Perhaps the prospect of better resource utilization will someday spark an interest in turning this landed forestry idea into reality. As our land base dwindles while our demands continue to expand, we may not be able to afford the wastefulness inherent in our current practice. In a society with a desperate need for wood fiber, it makes no sense to be cutting down the trees before they reach their peak productivity, but this is precisely what the timber companies are doing. It makes no sense to allow perfectly good growing land — the space between planted seedlings — to lie fallow for a decade or more. It makes no sense to destroy the biomass with herbicides, not utilizing it for energy production or other purposes. It makes no sense to plant pure stands of timber when mixed stands will grow more vigorously. Indeed, it makes no sense to jeopardize the health of tomorrow's timber by reducing a complex, but secure, forest ecosystem to a simple, but insecure, monoculture.

Furthermore, it makes no sense to rely exclusively on energy-consuming techniques of forest management at a time when energy is becoming increasingly scarce and dear. It makes no sense to replace men with machines at a time when high unemployment is becoming a built-in feature of our economy. It makes no sense to ignore the contributions that industrious, hard-working citizens could make for the sake of the future forests.

The timber companies, of course, have different ideas of what makes sense and what doesn't. For them, short-term financial considerations are paramount. Theirs is a world of interest rates, taxes, and bottom-line economics. Unfortunately, the economic structure of the capitalist market is ill suited to the practice of silviculture. In order to return a quick profit on their investment, timber companies short-circuit high interest rates by cutting adolescent trees. They destroy noxious weeds and simplify the forest ecosystem so it can be dealt with mechanically. To avoid high labor costs, they take the human presence out of the woods.

The public sector, however, is unblinded by the profit motive. The government can allow itself the freedom to maximize production instead of profit, to maintain a healthy and stable forest ecosystem, and to think of people as something more than "high labor costs." In fact, governmental policies are often quite admirable. The stated goal of the Forest Service is to practice multiple-use, sustained-yield forestry. This means that timber is to be treated as only one of the several resources of the forest, and that it is to be extracted only as fast as it can be expected to grow back. Other Forest Service policies try to encourage community development, seek citizen involvement, and give aid to underprivileged people. As a public agency, the Forest Service nominally recognizes the fact that silviculture cannot be practiced in a social vacuum, that decisions which affect the forest will affect the people as well.

These official policies are potentially significant, for the government plays a leading role in our forest management. There are not many sectors of the American economy that have been nationalized. That's not our style. Yet strangely enough, the government happens to own a considerable portion of the means of production for the forest products industry: over a quarter of the wooded land, and over half of the standing sawtimber. The government, therefore, has the power to effect meaningful changes in the practice of silviculture in America. The companies have a good excuse for failing to take the true interests of the forests and the people to heart: their primary constituency is the stockholders, not the trees or the workers. But the government has no such excuse. The "stockholders" of the government are the people — and the people have a vested interest in tomorrow's timber.

Let us suppose, then, that the government — and in particular the Forest Service, which manages the bulk of the wooded land in the public sector — takes its own stated goals to heart and decides to reexamine its basic silvicultural practices. Let us suppose that it establishes three basic categories for its forests.

(1) Some land is too remote, too sensitive, or simply too beautiful to be logged. These areas would be classified as wilderness. I won't argue here about whether we should have more wilderness or less. That's a separate issue. But there can be little argument that the earth ought to retain at least *some* land that is not significantly altered by human hands.

(2) Some land is too steep and sensitive to sustain a repeated human presence, but, if logged correctly, this land might still be able to support further generations of trees. These are the areas in which roadless logging systems such as skyline cables or helium balloons might be applied. Again, I could argue about how much land should be included in this category, but I won't. Suffice it to say that special care must be taken not to remove timber if the forest will not be able to stage a return, or if the land itself will be damaged beyond repair.

(3) Some land, probably most of the land, is capable of sustaining a repeated human presence and of producing commercial timber for the indefinite future, providing, of course, that the land is treated with care and respect. These are the areas in which landed forestry could be practiced. The land would be intensively managed according to the principles of holistic forestry. Landed foresters — not absentee managers or piecemeal workers — would caretake the land in such a way as to maximize the health and productivity of the future forest. The public land would no longer be treated as adjunct to the timber industry. The government would set the pace in establishing a social structure to encourage, rather than inhibit, the practice of holistic forestry.

Does that sound too idealistic? Is it impossible to conceive of a giant federal bureaucracy so drastically altering its basic *modus operandi?* Unlikely, perhaps — but not impossible. True, the cards are stacked against it. Vested interests of all sorts would be threatened by switching over to a system of landed forestry. Bureaucrats would lose their jobs. Of course, they could be retrained, but then they'd have to forsake the safe confines of their offices in favor of the rigors and uncertainties of working in the woods. Logging contractors, too, might be out of work, for the harvesting would be done by the workers who live on the land. Planting contractors, helicopter pilots, chemical manufacturers — the list of those affected by such a basic change in policy goes on and on. And each of these groups, quite naturally, could be expected to oppose the landed foresters program.

But isn't this true of any type of socioeconomic transformation? People are constantly having to readjust their jobs and their professions to meet the needs of the changing times. And one need of *these* changing times is to make our forests into healthy, stable environments that can supply us with timber indefinitely. Ultimately, the jobs created by a landed foresters program would be more numerous and more satisfying than the jobs that the program would replace. But the problems of transition remain; the inertia of the status quo is often overwhelming.

Then there's the problem of ideology. There's something a bit too *pure* about the landed forester idea. It's just not the American Way. We are used to being paid for exploiting resources, not enhancing them. We are used to selling our products and services to the highest bidder. We are used to fulfilling our contractual obligations with a minimum of effort and at the least possible cost to ourselves, and then leaving the job behind to enjoy our financial rewards. In short, we are used to performing alienated, but well-defined, labor. And we don't seem to mind that much, so long as we are adequately recompensed.

Yet there are aspects of the landed forestry idea that appeal to the American self-image. A landed forester would get paid only for what he actually produced. He could not hitch a ride on a pork-barrel paycheck. He would have a financial incentive for performing hard, physical labor. The better he worked, the more he would produce, and the higher he would be paid. Having no boss in his daily chores, he would have to answer only to himself for the quality of the work he performed. He would be a self-sufficient jack-of-all-trades, who could fall a tree, fix a tractor, build a house. The landed forester, like the mythical pioneer of the past, would seek to control his own destiny.

The status quo is changed by economic, social, and political forces — not by a bright idea. But let us suppose for the moment that the increased demand for timber and wood energy triggers some basic self-questioning within the ranks of the Forest Service. Let us suppose that even-age monoculture begins to show signs of distress here in America, just as it has in Europe. Let us further suppose that the cost of fuel continues to double every few years, and the social costs of unemployment also continue to rise. Meanwhile, experiments with resource cooperatives among private landowners tentatively demonstrate the viability of holistic alternatives. At some point the powers that be in the Forest Service, nudged along by a growing movement of holistic forestry advocates, decide that something like the landed forestry program might actually be worth a try. In the public's behalf and for the sake of tomorrow's timber, they decide to give it a whirl.

KURT HANSEN
LANDED FORESTER OF THE FUTURE

The year is 2021. Fifty-eight-year-old Kurt Hansen has been a landed forester for most of his working life. His grandfather was a gyppo logger from Oregon; his father ran a tree nursery for Weyerhaeuser; and his mother also worked for Weyerhaeuser, as a chemist specializing in the commercial uses of wood residue. True to his family tradition, Kurt chose to work within the timber industry. As a youth he pieced together odd jobs as a tree planter and a choker setter. Then in 1987, the Forest Service announced the inception of an experimental landed forestry program. Kurt applied for a position and was accepted into the three-year training program. In 1990, Kurt was placed on one of the first ten landed forestry tracts in the country. Along with five other graduates of the training course, he was granted a stewardship on 1,900 acres of cutover government land in the foothills of the Cascades.

"We came to the Elk Creek tract with a used D-6 cat, a backhoe/loader, and a horse team. We also had a mobile mill just for our own use around here. That was it. That and three years of training. We were supposed to harvest 400,000 board feet of sawtimber — and about half that much wood fiber — in our first year.

Today, our quota is twice that much, and we have no problem at all in reaching it. But it wasn't so easy back then. The land wasn't in the best of shape. There wasn't all that much residual timber, and a lot of the initial plantings had failed. There just weren't enough trees growing on the place. They should have let us take a few years to get things together and restock it all before giving us a harvesting quota.

"Of course, I understand why they made us harvest from the start. We had already been on the public payroll for three years at training school. The program was experimental. They had to show them we could pay off the capital expenses on the equipment. And the only way to do that was to harvest some trees.

"They gave us a flat salary for the first five years. The harvesting paid for our salaries, our operating expenses, and our payments on the equipment. But that was all. There would be no 'profit' for the government until we had enough time to play nursemaid to the land.

"Then after five years we got our first inventory report. A busload of students and teachers came out and camped on the land for two or three months. They measured everything they could get hold of, even the suppressed trees that didn't count in the final inventory. They came up with some pretty good numbers. Our salaries were jacked up quite a bit for the next five-year period.

"But our harvesting quota was jacked up, too. The funny thing was, the building market was depressed at the time, and here we were harvesting more timber. I was raised with the idea that the number of housing starts determined how many trees you should cut. It was hard to get used to the Timber Stabilization System. But now that we have Stabilization, it's hard to see how we got by without it. How can you practice real silviculture, how can you plan for the future, unless you know how many trees you're going to have to cut? You can't cut a hundred thousand feet this year, a million feet next year, and then nothing the year after. You need to get a rhythm going. And your rhythm has to approximate something that the land feels comfortable with. So that's why the government buys all our timber first, then releases it on the market when the demand calls for it. It's just like wheat and corn.

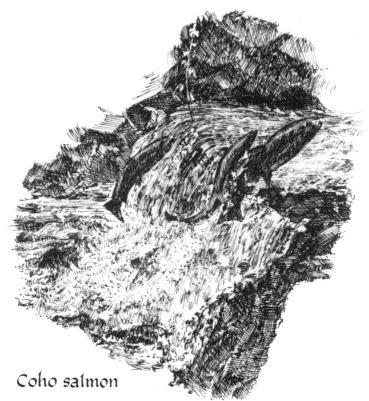

Coho salmon

The government stabilized agricultural crops several decades before it got around to stabilizing timber.

"You've got to have some kind of consistency if you're going to practice sustained-yield silviculture. That's what this Landed Foresters Program is all about. Everything's done one step at a time, day after day, year after year. We've been doing the same jobs now for thirty years. Fall trees, buck them up, snake them out. Chip the slash, truck it away. Plant the seedlings, trim the brush. Harvest the alders and the poles with horses. A horse only takes one step at a time. So do we. But a horse needs direction; it can't just wander about the woods, or the poles will never get to the landing. And we need direction, too. We need to know that we're raising so many trees to harvest in ten years, so many to harvest in twenty years, and so on.

"That's the craziest part of this incentive system: you're always thinking about the future. You release a six-foot Douglas-fir from its competition, and you don't see a six-foot sapling, you see the twenty-foot tree that they'll count on your next inventory, or maybe the hundred-foot tree that'll be harvested in a few decades. You work in the present, but you're always working toward the future. But the future never comes. The forest just keeps growing.

"Well, it doesn't *always* keep growing. Like this beetle infestation we had ten years ago. It wasn't our fault. It didn't start on our land. Actually, it was nobody's fault. It just *happened*. But *we* had to pay the consequences. We had to cut over three million feet of timber that year to stop the beetles. All our other work was neglected. We applied to the Disaster Bureau for compensation. They said it wasn't big enough to be called a 'disaster.' They counted some of it off against our harvesting quotas for the next few years, but that still didn't help our inventory situation. It set us back a lot in our thinning and brushing. We had to postpone a major logjam removal, so we lost an important spawning bed for that season. When the next inventory report came around, we didn't fare too well. Indirectly, we lost a lot by that beetle fiasco. But what can you do? It's not called a 'disaster' unless it wipes out over 10% of your stocking. So I guess it was just a 'misfortune.'

"Who decides those things? Don't let anyone ever tell you that there's no bureaucracy in the Landed Foresters Program. There's the Disaster Bureau, the Training and Inventory Department, the Sales and Stabilization System. Then there are representatives from each of these in the Harvesting Quota Division. Sure, qualified Landed Foresters account for a majority in all departments. But so what? It still feels like a bureaucracy to me. The mathematics of it is way over my head. I'll never serve in one of those departments, although I am thinking of putting in a stint as a training teacher. They send out questionnaires to get our input. They ask, 'Should timber inventories count for 45%, 55%, or 65% of the incentive schedule?' Or, 'Have you experienced any problems with the official fish counts not coinciding with your own estimates?' Sure, I've experienced problems with that. So I write it down. But who reads it? who actually changes the rules? That's what I call bureaucracy. That's politics. You can't escape it.

"Don't get me wrong. I think the system is basically as fair as it can be. It'll never be perfect, that's all. Take the inventories, for instance. One group of students measures things one way, and then five years later another group of students with a different teacher does it slightly differently. The differences might be small, but they can add up.

"A few years ago, enough of us Landed Foresters complained about this, so they made an experiment. They sent two different groups to evaluate the same tract. There was a 6% variation in the timber growth estimates and a 13% variation in the fisheries estimates. Some of the differences balanced each other out, but altogether the Landed Foresters would receive 4% more money from one inventory report than they would from the other. So, like I say, nothing's perfect.

"Of course, you try not to get petty about these things, but sometimes it's hard to resist the temptation. You visit another Landed Forestry tract, and you can't help comparing the virtues of their land with your own. You say, 'Boy, if I only had a layout like this, I could produce the best damn timber in the state.' But then you look at their inventory quota and incentive schedule, and you see that they're *expected* to produce the best damn timber in the state. Then you feel relieved that you don't have the extra tension of living up to such high expectations.

"Ultimately, you're always competing against yourself, not against the other tracts. The harvesting quotas and incentive schedules are personalized for each separate unit. But you still feel the effect of competition, even if it's only against yourself. You're always wanting to do it better this year than last year, and you certainly don't want to do it any worse.

"That time when we received a poor inventory report, that was emotionally hard on all of us; not to mention that we had to accept a fairly significant salary cut for the next five years. Sure, we tried to blame it on the beetles or the Disaster Bureau, but scapegoating didn't always work. We could never totally escape the nagging self-doubts: had we really messed up? How come even our water quality reports were worse that time? Weren't we working hard enough? Were we slowly getting too soft, too sedentary? Or maybe just downright lazy?

"There are lots of times when you long for just a good, flat salary with no strings attached. You get tired of having incentives to make you feel guilty for not working hard enough. Sometimes I'd rather be a plain, ordinary worker who just takes orders and then gets to go home and not worry about a thing. Sometimes I wish I had a boss to get mad at behind his back. Sometimes the responsibility of it all just wears me out. It makes me sick. Sick and tired.

"That's on my bad days, of course. Then the very next day I might feel suddenly inspired. Why, I don't know. One day, I'll be planting trees and grumbling to myself, and the following day the whole picture is revealed to me in all its splendor: this piece of the earth which I am so fortunate to inhabit, the productivity of the land, the worthwhile nature of my job. I see myself nourishing the land, helping the forest to grow. I see it all fit together: the trees, the fish, the deer, the water . . . even the hikers passing through. And I'm glad to be a part of it — an *important* part. I feel like I have a mission. My life has a purpose.

"Whether I'm an inspired missionary or an anxiety-ridden entrepreneur, I still do my job. I guess that's why the Landed Foresters Program has been working out so well: it appeals to both the best and the worst in us. Whether I'm motivated by guilt, greed, God, or glory, there's always a reason for me to get off my butt and get to work. I work on public land, I'm a public steward — and let me assure you, the public is getting its money's worth out of me. I'm out on the job ten, twelve hours a day, six days a week, rain or shine. Of course, I'm not always breaking my back all that time. Sometimes I just walk around, checking things out. I combine business with pleasure. If I need to relax, I'll decide it's a good time to read some water gauges down at the southern boundary or maybe look over the new plantings up on the ridges for signs of animal damage. I can take the whole afternoon off that way and still be doing my job.

"One problem with that, though, is I'm responsible not only to myself, but also to the rest of the Elk Creek team. We all like to check out the far reaches of our tract. The longer we have to walk, the better. So you don't really take off just any old time. You have to work it out with the others. And a lot of times, nobody takes off because we're all working together on the same project. Harvesting timber is that way. It goes a lot better if we're all pitching in. Removing logjams, too. Sometimes we all like to work together even if we don't have to, just so we can see some immediate results. When you're planting trees, for instance, it's good for morale. You can really cover a lot of ground when everybody's working at once.

"But at other times it seems we should get paid extra for just having to get along with each other. For the first few years, we used to have regular meetings once a week, which wasn't so bad, except that some of the meetings would last all night and then we'd all be worthless the next day. It was just one thing after the other. Should anyone except Sam ever run the cat? If so, would Sam be left to put together the pieces if the cat broke down? But if nobody else ever ran it, we'd all forget our training. Was it really just Sam's personal machine, purchased at group expense?

"That little hassle went on for about a year. Finally we got tired of it and just let things happen, which meant that Sam was our regular catskinner, until he had to miss a whole season because his mother was dying. So Nathan took over and didn't do such a bad job after all. Fact is, he was so bloody careful that the thing never quit on him once. He ran that machine as if Sam was looking over his shoulder every minute. The only problem was that he was a nervous wreck by the end of the season, so he was perfectly happy to have Sam take over as soon as he came back.

"Then there was the women's question. I guess it was a men's question, too. The problem was whether the women would always do the same job as the men. Ideologically, we wanted to be completely equal, and we thought that meant doing everything that everybody else did. But when we got right down to the nitty-gritty, we found that my wife Katie just doesn't like to use a chain saw. She likes to plant, she'll set chokers, she'll do anything but cut. On the other hand, she's great with the horses and she knows ten times as much as anybody else about hydrology. So we finally let that one ride, too. We relaxed our ideology and let Katie do the things she does best. She develops all the springs and she's our own personal expert about what might cause erosion. She tends the fish-rearing ponds. She drives the horse teams. That's fine by me, because when you get right down to it, I feel the same way about horses as she does about chain saws: I'm scared of them. I'm basically a cutter, a faller. Sam's a catskinner and a mechanic. Katie's a horse driver. So we developed a division of labor after all. Sure, each of us does lots of different jobs. But we each have our specialties, too.

"One of the biggest problems we ever had to work out was what happened when Sam got married. There were six of us at the start, two married couples and two single men. We all went to training school together. We were tight. We applied for the Elk Creek tract as a group. When we got here, we all worked together as professional equals. So we split our salaries evenly, right down the line.

"But when Sam's wife came out here, she wasn't a professional worker. She wasn't on the payroll. She chipped in here and there, but woods work isn't her natural calling. On the other hand, she couldn't get a job in town either, because we're just too far away for a daily commute. So the two of them had to live on a single salary, which was hard; plus it wasn't very fair, because, in fact, Karen did a lot of odd jobs around the place. She picked up a lot of loose ends, like tutoring the kids, for instance. Finally, we just cut her in as an equal partner and called her the teacher, the gardener, and so on. And she still helps out in the woods to some extent, mostly on planting days. It only made sense to cut her in, but it did mean less money for the rest of us.

"When you get right down to it, that's the touchiest part of this Landed Foresters Program: what happens when people want to change their lives? Karen moved in. O.K., we worked that one out. But then Dwight said he'd had enough after ten years. He was still single and not meeting any women out here. He wanted to go back to school, so he left. They sent us a replacement, fresh out of the training school. This replacement didn't know us and we didn't know him, yet we were supposed to become like a family. Fortunately, the new fellow worked out. If he hadn't, they would have sent out someone else. There is some flexibility in it.

"But that didn't solve Dwight's problem. He had to go out in the world and start from scratch. He'd been working all these years, but he had no equity. We don't own the place, so he had nothing to sell. Sure, we all pitched in and gave him a sort of termination paycheck, so he'd have something to work with. But he should've gotten more. The program should work something out for guys like that.

"We do have a retirement system, but it doesn't help if you've only worked ten years. The retirement plan is nice, though. You get a regular government pension, and you also get to stay on your land. If you're sixty years old and you've been working there for fifteen years or more, you get to stay there. They figure it's pretty much your home, even if you don't own it. But nobody knows how the retirement plan is really going to work, because there are only a few Landed Foresters who are old enough yet. Most of us were young when we joined. There are a lot of people around my age, around fifty or so, who are still with the program. So we'll see pretty soon. Are we going to want to stick around and grow old right here? Will we want to travel? Will we want to move closer in, so we can get better medical care? Nobody knows.

"I will say this about the Landed Foresters Program: they try to be flexible. Since most of the administrators are just on leave from their own tracts, they really understand the everyday problems. The money isn't the greatest; nobody gets rich in landed forestry. But the working conditions are appealing, particularly if you like to be outdoors. And it's a respected occupation. They've got huge waiting lists of people trying to crack into the training program. Maybe that's one reason that the companies are finally getting interested in landed forestry: they see all the good, young workers that are just knocking at the door, and they figure they might be able to get more work out of *them* than they get out of the contractors they use now. But landed forestry won't be the same when the companies use it. Their incentive schedule is going to be a lot more straightforward: timber growth rates, that's all that'll count, which is all right as far as it goes. At least they're recognizing that people can grow timber better than machines.

"The government, see, has forced them to recognize that. We're just now beginning to see the results of the holistic forestry methods used in the Landed Forestry system. The Landed Forestry tracts are producing more — and better — timber. It's not full of knots. It's taller and straighter, because the trees were grown within well-managed, mixed stands. The conifers had to shoot upward to keep pace with the alders and the hardwoods. They didn't waste all their energy on limbs. Then just at the right moment, the competition was removed, and the conifers were released to close their crowns. With the ground loosened and enriched, the conifers prospered. And now they're approaching maturity — some are about ready to harvest. Fact is, if they were on company land, they'd already *be* harvested, because the companies are cutting down stuff that's much smaller than what we still have standing.

"And the best of it is, the government has done it all at no extra expense. The alders and the hardwoods, they've paid their way out of the woods. And the extra labor . . . well, like I say, nobody's getting rich off the Landed Foresters Program. We get by, we have a good life. But living the way we do, we simply don't *need* the kind of money that the contractors do.

"We're not complaining. We're independent. We control our own destinies. We get paid for growing timber, for tending the forest, for caring about the future. You can't ask for more satisfying work."

Glossary

Agent Orange A combination of 2,4-D and 2,4,5-T used by the United States to defoliate forests in Vietnam during the 1960s.

A-Horizon Scientific term for the upper layer of soil.

All-Age Management (also, All-Aged Management) The management of a forest to include trees of diverse ages and, therefore, of diverse sizes.

Anadromous Fish Ocean fish that migrate up a river to spawn.

Angle of Repose The maximum slope at which a hillside will remain stable; a hill that is made steeper than its angle of repose is likely to give way to landslides.

Back Hoe A piece of earth-moving equipment with a trench-digging shovel on one end and a loading scoop on the other.

Bare-Root Seedlings Seedlings germinated together in a single mass of dirt instead of in their own separate containers.

Biomass The sum total of organic material in a given area.

BLM The Bureau of Land Management, an agency of the United States Department of the Interior.

Board Foot A unit of measurement nominally equal to a board measuring 1 in. x 12 in. x 12 in.

Boomer A migrant logger.

Broadcast Burn Controlled fire over the entire surface of a designated area.

Broadleaf Tree A tree with conventional leaves, not needles or scales.

Buck To cut a tree into logs after it has been felled.

Bucker A person who cuts felled trees into logs.

Bull Donkey A large steam donkey operating along skid roads.

Bull-Puncher (also, Bull-Whacker) The driver of a logging ox-team.

Butt Log The log taken from the base of a tree, often slightly irregular.

Calk Boots Logging boots with short spikes set in the soles.

Cat Abbreviation for caterpillar.

Caterpillar In common usage, a logging tractor. (In this text, the word "caterpillar" denotes any logging tractor; the word "Caterpillar" denotes a logging tractor made by the Caterpillar Company.)

Catskinner The operator of a logging tractor.

C.E.T.A. Comprehensive Employment and Training Act, a federal job-creating program.

Chain Grab A hook and chain used as an anchor for a cable network.

Chaser A person who works on the landing in a cable operation, unhooking logs from the chokers.

Check Dams A series of small dams in a stream or drainage ditch, used to minimize erosion.

Choker A cable loop that is attached to a log during the yarding of timber.

Choker Setter A person who puts chokers around logs.

Clearcutting The removal of the entire stand of trees within a designated area.

Climax The culminating stage of plant succession for a given environment, where the vegetation has reached a stable condition.

Codominant Trees Trees that reach into the upper forest canopy but do not extend above their neighbors.

Concave Slope A slope that appears slightly hollowed out, like a crescent moon.

Conifer A cone-bearing tree that generally has needles or scales. Most commercial tree species in the Northwest are conifers.

Conk Fruit-mass of a wood-destroying fungus, generally projecting from the trunk of a tree.

Contour Wattles A series of hillside terraces supported by a network of live shoots and intertwined dead branches.

Convex Slope A slope that appears bulged or pushed outward; the opposite of concave.

Cord A unit of measurement for firewood or pulpwood, equal to a pile measuring 4 ft. x 4 ft. x 8 ft., or 128 cubic feet.

Corks See *Calk Boots.*

Crawler Tractor A logging tractor with ribbed metal tracks instead of wheels.

Cruise A survey to measure the timber in a designated area.

Cull A tree or log that is rejected because of defects.

D.B.H. Diameter at breast height, the standard method for determining the size of a tree is to measure the diameter of the trunk at a point 4½ feet above ground level.

Deciduous Forest A forest composed primarily of trees that shed their leaves in the winter.

Deck A pile of logs; logs that are about to be transported or milled form a hot deck; logs that are being stored form a cold deck.

Dibble A pointed digging tool used for planting containerized seedlings.

Dog A hook.

Dog-hole Port A small, often tenuous, anchoring site along a rugged coastline.

Dominant Trees Trees that are taller than their immediate neighbors.

Donkey Puncher The operator of a steam donkey.

Ecosystem The complex of interactions between a community of living organisms and their environment.

E.I.R., E.I.S. Environmental Impact Report, Environmental Impact Statement, studies that evaluate the effects of a proposed action on the environment.

EPA Environmental Protection Agency, an arm of the federal government charged with evaluating the safety of commercially used chemicals.

Erosion Hazard Rating A number, based on the soil type and the steepness of the slope, which is supposed to determine a site's susceptibility to erosion.

Even-Age Management (also, Even-Aged Management) Forest management in which the crop trees are all of a similar age.

Fall (also, Fell) To cut down a tree.

Faller A person who cuts down trees.

Feller-Buncher A machine that cuts trees with giant shears and then stacks the trees in a pile.

F.I.P. Forestry Incentives Program; a federal cost-sharing program that helps small timber owners pay for reforestation or timber stand improvement.

Fry Young fish, just after they are hatched.

Girdling Killing a tree by removing a strip of bark from around its trunk.

Grapples Giant arms on a logging machine which can grab logs and move them from place to place.

Green Logging The logging of timber that is still alive; the opposite of salvage logging.

Green-Manure Trees Trees that fertilize the ground as they grow, generally by the presence of nitrogen-fixing bacteria living on their roots.

Guy Line A support cable for a spar tree or yarding tower.

Gyppo Logger A small, independent contractor.

Hand Briar See *Misery Whip*.

Hardwood Popular term for a broadleaf tree.

Haulback Line The cable that returns the chokers to the woods after the yarding of a turn of logs.

Haul Road A major road engineered for logging trucks, as well as tractors.

Heartwood The inner layers of a tree which have ceased to contain living cells and which serve only for structural support; lumber coming from heartwood is generally stronger and rots less quickly than that coming from the outer layers of the tree (sapwood).

Herbicide A chemical used to control unwanted vegetation.

High-Grading Logging in which the high-quality timber is removed while the low-quality timber is left standing.

High-Lead Cable logging in which the main block is suspended from a spar tree or yarding tower so that the front end of the log can be lifted off the ground.

Hoedad A digging tool used for planting bare-root seedlings.

Hog Fuel (also, Hogged Fuel) Sawdust, shavings, and other wood wastes used to power industrial boilers.

Holistic Forestry A system of forestry which treats the ecosystem as a self-equilibrating mechanism and which proposes only minimal, site-specific interference with natural processes.

Hooker In modern usage, a person who attaches the tag lines dangling from a helicopter to the chokers attached to the logs; in the old days, another word for hooktender.

Hooktender Woods boss, or foreman of a yarding crew.

Intensive Management The entire complex of industrial tree-farming techniques: genetic selection, site preparation, planting, weed control, pest control, fertilization, thinning, harvesting.

Landing A place where logs are assembled for loading.

Leader The top of a tree, representing the most recent growth.

Mainline The cable that moves the logs in a yarding operation.

Manual Release Hand-clearing of brush (as contrasted with the use of herbicides) to lessen the competition around desired trees.

Mean Annual Increment (also, M.A.I.) The average annual growth rate for a tree, computed over its entire life cycle.

Mensuration The branch of forestry concerned with the measurement of timber.

Millrace The controlled channel of water leading up to a mill.

Misery Whip A large, hand-operated crosscut saw used for falling and bucking timber.

Monoculture The raising of a tree crop that consists of a single species.

Multiple Use Management that encourages several distinct uses of the forest.

Mycorrhizae Fungi that grow among the outer cells of plant rootlets and form a symbiotic relationship with their hosts. The fungi receive photosynthetic products such as carbohydrates and vitamins from the host plants; in return, they help their hosts absorb nutrients and water from the soil, and they often give protection against disease.

Nitrogen Fixation The transformation of atmospheric nitrogen into nitrogen compounds that can be used by growing plants.

Overstory Removal The harvesting of the tallest trees without purposely removing the smaller vegetation.

Overtop To grow above the neighboring vegetation.

Peavy A wooden pole, with a spike at the end and a hinged hook, used for moving and turning logs by hand.

Peeler A log of sufficient size and quality for making rotary-cut veneer.

Phenoxy Herbicides A group of closely related chemical compounds (most commonly 2,4-D; 2,4,5-T; and 2,4,5-TP) which can kill, or severely damage, many species of broadleaf plants by promoting the uncontrolled expansion and division of cells.

Photosynthesis A green plant's transformation of carbon dioxide and water into carbohydrates, using sunlight as the source of energy.

Pioneer A plant belonging to the earliest stage of vegetational succession; pioneers establish themselves quickly in the wake of cataclysmic changes and then are gradually replaced by successor species.

Plus Tree A tree thought to be of superior genetic quality.

Productivity Class See *Site Class.*

Release Freeing the crop tree from immediate competition by eliminating, or retarding the growth of, the neighboring vegetation.

Riffle A shallow rapid in a stream.

Rigging The cable network, including blocks and related hardware, used in yarding timber or in directional falling.

Rigging Slinger A person who attaches chokers to the main yarding cable.

River Rat (also, River Hog) A person who follows the logs on a river run and pries them loose when they cease to move.

Rotation The number of years allotted to a single tree crop in a given location.

Salvage Logging Logging in which only dead timber (either standing or on the ground) is removed; the opposite of green logging.

Sapwood The outer layers of a tree which contain living cells; lumber coming from sapwood is generally weaker and rots more quickly than that coming from the inner layers of the tree (heartwood).

Scale To measure harvested logs.

Second Growth Young trees that grow in the wake of a prior harvest.

Seed Tree A tree that is purposely left standing during a logging operation in order to produce natural regeneration in the surrounding area.

Selective Cutting The periodic harvesting of selected trees scattered throughout a forest, either individually or in small groups.

Shelterwood Cutting The harvesting of timber in two or more successive stages, where the trees left standing after the first harvest provide a natural seed source and partial protection for regeneration.

Show A logging operation.

Side A complete yarding and loading crew.

Side-Spooler A single-engine, single-spool steam donkey.

Silvex The phenoxy herbicide 2,4,5-TP.

Silviculture The cultivation of forest trees.

Site Class A grouping of similar site indices.

Site Index A measure of the productive capacity of a given area, based on the height of the dominant trees at a specified age; a site index of 170 for Douglas-fir means that a 100-year-old dominant tree in that stand will be about 170 feet tall.

Site Preparation Preparing an area for planting, often with the aid of controlled fires, herbicides, or mechanized devices.

Skid To drag logs along the ground from the forest to a landing.

Skid Road A road used for skidding timber. In the old days, skid roads were made of wood; today, a logging tractor carves its own skid roads in the dirt before removing the timber.

Skid Trail A path through the woods created by the skidding of logs.

Skyline A cable logging system in which the logs are suspended in air rather than dragged along the ground.

Slacker An early term for a skyline cable.

Softwood Popular term for conifer.

Spar Tree A tree that has been limbed, topped, and rigged with cables to provide elevation in high-lead yarding.

Splash Dam A temporary dam on a shallow stream. When the water behind the dam is released, logs can be propelled downstream.

Springboard A board inserted into a notch in the trunk of a large tree, forming a one-piece scaffolding upon which a faller can stand while he cuts down the tree.

Stand A community of trees managed as a collective unit.

Steam Donkey A steam-driven yarding machine used in the late nineteenth and early twentieth centuries.

Stocking The quantity of crop trees in a designated area, often expressed as a percent of maximum capacity.

Strip-Cutting The harvesting of parallel strips of timber, in which the areas between those strips are left untouched.

Stumpage The quantity or value of standing timber.

Succession The replacement of one plant community by another as local conditions change over time.

Successor Species Species of plants that thrive in the intermediate stages of forest succession, after the pioneers, but before the climax stage.

Sunscald Damage done to a tree because of sudden and excessive exposure to the sun.

Super Trees The timber industry's term for fast-growing, high-yield trees created by genetic engineering and aided by intensive forest management.

Suppressed Tree A tree that remains in the forest understory because its growth is inhibited by neighboring vegetation.

Sustained Yield The amount of timber that a forest can produce on a continuing basis in perpetuity, determined by a careful management program.

Swing Donkey A steam donkey used to supplement another steam donkey over a long haul.

Tag Line A short stretch of cable that connects the chokers and logs with the main part of the yarding system.

Tailhold The anchor on the far end of a cable yarding system.

T.H.P. Timber Harvest Plan; a document that states how a timber harvest will be executed; by law, a T.H.P. must be filed with the State Department of Forestry prior to any logging operation in California.

Timber Stand Improvement The pruning, weeding, and/or thinning of a stand of timber.

Tolerance According to most foresters, the ability of a tree to grow in the presence of shade and in competition with other trees; according to some foresters, the ability of a tree to grow under any adverse conditions.

Transpiration The process by which plants release water into the atmosphere.

Turn of Logs A group of logs yarded at the same time by the same machine or animal.

Uneven-Age Management (also, Uneven-Aged Management) see *All-Age Management.*

Water Bar A mound of dirt placed in a logging road to direct water away from the roadbed, thereby lessening the damage from erosion.

Waterslinger A person who attaches logs to a donkey-powered cable along a wooden skid road and then lubricates the skid road as the logs are dragged to the landing.

Whistle Punk A person who communicates signals to the various workers on a steam-donkey yarding crew.

White Fingers Ramos' Disease; a contraction of the blood vessels in the fingers due to excessive vibrations over an extended period of time, causing discoloration, fatigue, and loss of feeling; a common ailment among tree fallers, buckers, and brush cutters.

Widow Maker A high, loose branch that presents a danger to loggers working beneath it.

Windthrow (also, Windfall or Blowdown) The destruction of standing timber by the wind.

Wobblies Members of the Industrial Workers of the World, a radical labor organization during the first quarter of the twentieth century.

Yarder A machine used in yarding timber.

Yarding The movement of logs from the woods to a central loading area, or landing.

Yield Table A table that attempts to predict how much timber of a given species could be grown in a perfectly stocked stand with a specified site index.

Notes

CHAPTER 1

1. "Reminiscences of Mendocino," *Hutchings California Magazine,* October 1858.
2. Ron Finne, *Natural Timber Country* (16 mm film).
3. Jess Larison 1979: personal communication.
4. Gifford Pinchot, *The Fight for Conservation* (Seattle: Univ. of Washington Press, 1967), pp. 14-15.
5. 16 *United States Code* 475.
6. Shirley W. Allen and Grant W. Sharpe, *Introduction to American Forestry* (New York: McGraw-Hill, 1960), p. 331.
7. Finne, *Timber Country.*
8. Jess Larison 1979: personal communication.
9. U.S.D.A. Forest Service, Pacific Northwest Forest and Range Experiment Station, *Douglas-fir Supply Study* (Portland, Oreg., 1969).
10. Finne, *Timber Country.*
11. Ibid.

CHAPTER 2

1. Herman H. Chapman and Walter H. Meyer, *Forest Valuation* (New York: McGraw-Hill, 1947), p. 88.
2. Georgia-Pacific Corp., *1977 Annual Report to Shareholders.*
3. *American Forests,* October 1978, p. 15.
4. Nancy Wood, *Clearcut: The Deforestation of America* (San Francisco: The Sierra Club, 1971), p. 17.
5. Gifford Pinchot, *Breaking New Ground* (New York: Harcourt, Brace, 1947). Cited in Daniel R. Barney, *The Last Stand: Ralph Nader's Study Group Report on the National Forests* (New York: Grossman, 1974), pp. 64-65.
6. Wood, *Clearcut,* p. 30.
7. Ibid., p. 62.
8. Ibid., p. 71
9. Philip F. Hahn, "The Effect of Animal Damage on Volume Growth," (Eugene, Oreg.: Georgia-Pacific Corp., 1978).

10. *Report from the California Redwood Association,* (Eureka, Calif.: Redwood Region Logging Conference, 1978).
11. *Forest Industries,* November 1978, p. 60.

CHAPTER 3

1. Finne, *Timber Country.*
2. Esteban de la Puente 1980: personal communication.
3. R. W. Stark, "The Entomological Consequences of Even-Age Management," *Even-Age Management,* ed. Richard K. Hermann and Denis P. Lavender (Corvallis, Oreg., 1973).
4. John R. Parmeter, Jr., "Ecological Considerations in Even-Age Management: Microbiology and Pathology," *Even-Age Management,* ed. Hermann and Lavender, pp. 122-123.
5. Samuel A. Graham, *Forest Entomology* (New York: Reinhold, 1952), p. 66.
6. Frederick E. Smith, "Ecological Demand and Environmental Response," *Journal of Forestry,* December 1970, p. 755.
7. Roy R. Silen, "The Care and Handling of the Forest Gene Pool," *Pacific Search,* June 1976, p. 9.
8. R. E. Martin, "Prescribed Burning for Site Preparation in the Inland Northwest," *Tree Planting in the Inland Northwest* (Wash. State Univ. Conf., 1976).
9. Robert F. Tarrant, K. C. Lu, W. B. Bollen, and J. F. Franklin, *Nitrogen Enrichment of Two Forest Ecosystems by Red Alder,* Forest Service Research Paper PNW-76 (Portland, Oreg., 1969).
10. Robert F. Tarrant, K. C. Lu, W. B. Bollen, and C. S. Chen, *Nutrient Cycling by Throughfall and Stemflow Precipitation in Three Coastal Oregon Forest Types,* Forest Service Research Paper PNW-54 (Portland, Oreg., 1968).
11. Tarrant, Lu, Bollen, and Franklin, *Nitrogen Enrichment.*
12. Joseph Kittredge, *Forest Influences* (New York: McGraw-Hill, 1948).
13. D. D. Harris, "Hydrologic Changes after Clearcut Logging in a Small Coastal Oregon Watershed," *J. Res. U.S. Geol. Survey* 1 (4): 487-491. Cited in Marvin Dodge, L. T. Burcham, Susan Goldhaber, Bryan McCulley, and Charles Springer, *An Investigation of Soil Characteristics and Erosion Rates on California Forest Lands* (Sacramento, Calif., 1976), p. 25.
14. Gerald Myers 1979: personal communication.
15. W. D. Ellison, "Soil Erosion," *Soil Sci. Soc. Amer. Proc.* 12 (1947): 479-484. Cited in Dodge, et al., *Erosion Rates,* p. 13.
16. Robert N. Coats, "The Road to Erosion: A Cautionary Tale," *Environment,* January-February 1978.
17. R. L. Fredriksen, *Erosion and Sedimentation Following Road Construction and Timber Harvest on Unstable Soils in Three Small West-*

ern Oregon Watersheds, Forest Service Research Paper PNW-104 (Portland, Oreg., 1970).

18. C. T. Dyrness, *Mass Soil Movements in the H. J. Andrews Experimental Forest,* Forest Service Research Paper PNW-42 (Portland, Oreg., 1967).
19. Jack S. Rothacher and Thomas B. Glazebrook, "Flood Damage in the National Forests of Region 6" (Portland, Oreg., 1968). Cited in Dodge, et al., *Erosion Rates,* p. 39.
20. C. T. Youngberg, "The Influence of Soil Conditions Following Tractor Logging on the Growth of Douglas-fir Seedlings," *Soil Sci. Soc. Amer. Proc.* 23 (1959): 76-78.
21. Henry A. Froehlich, "The Impact of Even-Age Forest Management on Physical Properties of Soil," *Even-Age Management,* ed. Hermann and Lavender, pp. 199-220.
22. Robert R. Curry, "Geologic and Hydrologic Effects of Even-Age Management on Productivity of Forest Soils, Particularly in the Douglas-fir Region," *Even-Age Management,* ed. Hermann and Lavender, pp. 137-178.
23. E. C. Steinbrenner and S. P. Gessel, "The Effect of Tractor Logging on Physical Properties of Some Forest Soils in Southwestern Washington," *Soil Sci. Soc. Amer. Proc.* 19 (1955): 372-376.
24. Ibid.
25. Curry, "Geologic and Hydrologic Effects."
26. Ibid.
27. Coats, "Road to Erosion."
28. S. A. Wilde, *Forest Soil and Forest Growth* (Waltham, Mass., 1946), p. 158.
29. Coats, "Road to Erosion."
30. Steinbrenner and Gessel, "Effect of Tractor Logging."
31. J. D. Hall and R. L. Lantz, "Effects of Logging on the Habitat of Coho Salmon and Cutthroat Trout in Coastal Streams," *Symposium on Salmon and Trout in Streams,* ed. T. G. Northcote (Vancouver, B.C., 1969).
32. R. L. Lantz, *Guidelines for Stream Protection in Logging Operations* (Portland, Oreg.: Oregon State Game Commission, 1971).
33. Coats, "Road to Erosion."
34. *An Environmental Tragedy* (Sacramento, Calif.: California Dept. of Fish and Game, 1971).
35. Coats, "Road to Erosion."
36. Wood, *Clearcut,* p. 70.
37. For a summary of these reports, with bibliographical references, see John W. Warnock and Jay Lewis, *The Other Face of 2,4-D* (Penticton, B.C.: South Okanagan Environmental Coalition, 1978), pp. II-1 to II-11.
38. San Francisco *Chronicle,* 9 April 1977.

39. Wilbur P. McNulty, Jr., "Testimony on the Effects of Tetra-Dioxin on Primates," Citizens Against Toxic Sprays v. U.S. Dept. of Agriculture, U.S. District Court, District of Oregon, 1976.
40. Warnock and Lewis, *Other Face*, pp. II-8 and II-9.
41. Ibid.
42. Ruthanne Cecil 1979: personal communication.
43. Terri Aaronson, "A Tour of Vietnam," *Environment*, March 1971; Philip Boffey, "Herbicides in Vietnam," *Science*, 8 January 1971; Hilary Rose and Steven Rose, "Chemical Spraying as Reported by Refugees from South Vietnam," *Science*, 25 August 1972; Ton That Tung, "Clinical Effects of the Massive and Continuous Use of Defoliants on Civilian Population," *Reunion Internationale de Scientifiques sur la Guerre Chimique au Viet Nam* (Paris, 1970); Ton That Tung, "Primary Liver Cancer in Vietnam," *Chirurgie*, 16 May 1973.
44. A. A. Bashirov, "The State of Health of Workers Manufacturing the Herbicide 2,4-D," *Vrachebnoe Delo* 10 (1969): 94-95. Abstract on file at the Environmental Protection Center, Fort Bragg, Calif.
45. Warnock and Lewis, *Other Face*, pp. VI-9 to VI-12.
46. J. Yoder, M. Watson, and W. W. Benson, "Lymphocyte Chromosome Analysis of Agricultural Workers during Extensive Occupational Exposure to Pesticides," *Mutation Research* 21 (1973): 335-340.
47. Associate Committee on Scientific Criteria for Environmental Quality, *Phenoxy Herbicides: Their Effects on Environmental Quality* (National Research Council of Canada, 1978).
48. Diane Courtney, *Prenatal Effects of Herbicides: Evaluation by the Prenatal Development Index*, 15th Annual Meeting of the Teratology Society, May 1975.
49. Philip C. Kearney, Edwin A. Woolson, and Charles P. Ellington, Jr., "Tetrachlorodibenzo-Dioxin in the Environment: Sources, Fate and Decontamination," *Environmental Health Perspectives* 5 (1973): 273.
50. R. Baughman and M. Meselson, "An Analytic Method for Detecting TCDD," *Environmental Health Perspectives* 5 (1973): 27-35; M. Meselson and P. W. O'Keefe, "TCDD Residues in Beef Fat and Bovine Milk," *The Biological Laboratories*, Harvard Univ., 12 December 1976); U.S. Environmental Protection Agency findings of TCDD residues in cattle and goats cited in Warnock and Lewis, *Other Face*, p. V-7.
51. E. Stanton Maxey, American Board of Surgery, to Russell E. Train, Environmental Protection Agency, 20 August 1974; M. Meselson and P. W. O'Keefe, "Human Milk Monitoring: Preliminary Results for Twenty-one Samples," *The Biological Laboratories*, Harvard Univ., 15 December 1976; Billie Shoecraft, *Sue the Bastards* (Phoenix, 1971); GHT Laboratories of Imperial Valley, Brawly, Calif.,

"Report on Phenoxy Herbicide Residues in the Blood of Oregon Residents of the Suislaw National Forest," December 1972.
52. Newspaper reports of these and other incidents on file at the Environmental Protection Center, Fort Bragg, Calif.
53. Jack Anderson, "Defoliating America," San Francisco *Chronicle*, 24 April 1978.

CHAPTER 4

1. Tarrant, et. al., *Nitrogen Enrichment*.
2. Robert F. Tarrant, "Stand Development and Soil Fertility in a Douglas-fir—Red Alder Plantation," *Forest Science*, September 1961.
3. Groundwork, Inc., *Willamette Brush Control Study Project* (Eugene, Oreg., 1978).
4. H. Gus Wahlgren, "Tapping the Forest Resource," *American Forests*, October 1978.
5. Ibid.
6. Norval Morey, "Clean the Land, Fuel the Country," *Logging Management*, June 1980, p. 24.

CHAPTER 5

1. Glen O. Klock, "Impact of Five Postfire Salvage Logging Systems on Soils and Vegetation," *J. Soil and Water Conservation*, March-April 1975.
2. Dodge, et al., *Erosion Rates*, p. 60.
3. Faye Stewart 1979: personal communication.
4. Steve Conway, *Logging Practices: Principles of Timber Harvesting Systems* (San Francisco, 1976).
5. Ibid; see also Doyle Burke, "Helicopter Logging: Advantages and Disadvantages Must Be Weighed," *Journal of Forestry*, September 1973.
6. Conway, *Logging Practices*.

CHAPTER 6

1. D. M. Moehring and I. K. Rawls, "Detrimental Effects of Wet Weather Logging," *Journal of Forestry*, March 1970.
2. P. M. McDonald, W. A. Atkinson, and D. O. Hall, "Logging Costs and Cutting Methods in Young Growth Ponderosa Pine in California," *Journal of Forestry*, February 1969.
3. U.S.D.A. Forest Service, *Final Environmental Statement and Renewable Resource Program, 1977-2020*, Report to Congress, 1976.
4. Ted Blackman, "Computer Tells Foresters How Specific Practices Would Work," *Forest Industries*, January 1979.

5. Promotional brochure: "Controlled Falling . . . A Lean Toward Higher Profits," Silvey Precision Chain Grinder Co., Eagle Point, Oreg.
6. Ibid; see also Dave Burwell, unpublished paper on controlled falling, Springfield, Oreg.
7. "The Uncommon Forest Management Plan," *Timber/West*, February 1978.
8. Ibid., p. 26.
9. Ibid.
10. Don Minore, "Shade Benefits Douglas-fir in Southwestern Oregon Cutover Area," *Tree Planters' Notes* 22 (1).

CHAPTER 7

1. Barney, *Last Stand*, pp. 143-144.
2. U.S.D.A. Forest Service, *The Outlook for Timber in the United States*, Report of the findings of the 1970 Timber Review. Cited in Barney, *Last Stand*, p. 34.
3. As of late 1980 Congress had just passed a law allowing a timber owner to amortize up to $10,000 in annual reforestation expenses over a seven-year period. This new tax break will certainly help the cause of reforestation. The law, however, is aimed at small holdings, since an expense of $10,000 per year is not significant in corporate economics.
4. At Weyerhaeuser's 1980 Annual Meeting, Senior Vice-President C. W. Bingham reported that his company spends approximately $110 million per year on "reforestation, forest management, research and development and holding costs including taxes." This represents about 3 percent of Weyerhaeuser's overall expenditures.
5. Site class II, site index 170.
6. Steven Calish, Roger D. Fight, and Dennis E. Teeguarden, "How do Nontimber Values Affect Douglas-fir Rotations?," *Journal of Forestry*, April 1978. "McArdle and Meyer's (1930) normal yield table for Douglas-fir, site index 170, was used as the basis for calculating a timber-only rotation. Yields, expressed in thousand cubic feet per acre, were reduced by a factor of 0.94 to account for deer browsing on early growth. For lack of better data, we assumed that rotations may be repeated without affecting site productivity. Stumpage revenue and initial costs were expressed in real dollars. Stumpage was valued from $700 per thousand cubic feet for 10-inch trees to about $875 for 21-inch trees. Regeneration cost was set at $75 per acre, and it was assumed that successful regeneration occurs immediately following harvest. For the data enumerated, and a 5 percent rate of interest, the best economic rotation for timber production is 36 years, resulting in an SEV (soil expectation value) of $643 per acre. In contrast, if culmination of mean annual increment is used, the rotation is 64 years and SEV is $329 per acre." (p. 218)

7. Site class II, site index 170. Yield table taken from T. F. Arvola, *California Forestry Handbook* (Sacramento, Calif., 1978), p. 217.

8. Ibid.

9. Site index 180; ibid., p. 215.

10. If stumpage prices rise significantly faster than the general rate of inflation, there will be an added incentive to wait. Today, stumpage prices are rising because most of the old-growth timber is gone, yet most of the modern tree farm plantations are still too small to be harvested. Most forecasts for the future, however, project that stumpage prices will level off, or even decline, as the tree farms planted in the last two decades reach maturity.

11. Yield table from Arvola, *Handbook,* p. 215.

12. For the exact amounts of log exports at any given time, see *Production, Prices, Employment, and Trade in Northwest Forest Industries,* published quarterly by the Forest Service, Pacific Northwest Forest and Range Experiment Station, Portland, Oreg.

13. Barney, *Last Stand,* pp. 143-144.

14. 16 *U.S.C.* 528-531. Excerpted in Barney, *Last Stand,* p. 145.

15. Gordon Robinson 1979: personal communication.

16. Barney, *Last Stand,* p. 8.

17. As of late 1980 Congress had just established a Reforestation Trust Fund to help eliminate the reforestation backlog in the National Forests. Funding will come from import duties on plywood and lumber and will terminate in five years.

18. Jack Shepherd, *The Forest Killers* (New York, 1975), p. 371.

19. Glen O. Robinson, *The Forest Service* (Baltimore: Johns Hopkins, 1975), p. 39.

20. Ibid., p. 34.

21. Samuel T. Dana and Evert W. Johnson, *Forestry Education in America: Today and Tomorrow* (Washington, D.C.: Society of American Foresters, 1963), p. 117.

22. Ibid., p. 142.

23. Ibid., p. 7.

24. Dodge, et al., *Erosion Rates,* pp. 83-88.

Selected Readings

GENERAL

Arvola, T. F. *California Forestry Handbook.* Sacramento, Calif., 1978. An excellent guide for small landowners interested in timber management, and a good start for laymen interested in a modest study of forestry. Included are a glossary and some commonly used tables. Write to:
> Office of Procurement, Publications Section
> P.O. Box 20191
> Sacramento, Calif. 95820

Forbes, R. D., ed. *Forestry Handbook.* New York: Ronald Press, 1955. The standard source of tables and technical information for practicing foresters, but of limited utility for laymen.

U.S.D.A. Forest Service. *The Forester's Almanac.* Portland, Oreg., 1977. A catalog of publications from Forest Service research in the Pacific Northwest. Includes brief abstracts and comprehensive bibliographies on all aspects of forestry. Write to:
> Pacific Northwest Forest and Range Experiment Station
> Forest Service, U.S.D.A.
> 809 N.E. Sixth Ave.
> Portland, Oreg. 97232

Forest Service publications listed below and marked "PNW" are available from this address. Also available are periodic lists of current publications, which are sent free-of-charge to any interested party. If you are interested in pursuing the academic aspects of forestry without going back to school, this ongoing bibliography will keep you up-to-date on the latest research in the field.

FORESTRY PERIODICALS

American Forests. Written in nontechnical language for a popular audience. Write to:
> The American Forestry Association
> 1319 Eighteenth St., N.W.
> Washington, D.C. 20036

Forestry Research West. Summaries of Forest Service research, also written for a popular audience. Write to:
>240 West Prospect St.
>Fort Collins, Colo. 80526

Forest Science. The papers here are highly technical. Write to:
>Society of American Foresters
>5400 Grosvenor Lane
>Washington, D.C. 20014

Journal of Forestry. A professional journal, but many articles are comprehensible to a layman. Write to:
>Society of American Foresters
>5400 Grosvenor Lane
>Washington, D.C. 20014

CHAPTER 1. TREE MINING: THE VOICE OF HISTORY

Andrews, R. W. *This Was Logging!* Seattle: Superior Publishing Co., 1954. A whole genre of big, fancy picture books glorify logging in the old days. Darius Kinsey's photos make this one the best of the lot.

Cox, Thomas R. *Mills and Markets: A History of the Pacific Coast Lumber Industry to 1900.* Seattle: Univ. of Washington Press, 1974. A careful and detailed economic history of logging in the Northwest.

Finne, Ron. *Natural Timber Country.* 16 mm film. A vivid, accurate, and humane picture of logging in the old days. Contains much historical material available from no other source. One hour long. To rent or purchase a print, write to:
>Ron Finne
>36526 Jasper Rd.
>Springfield, Oreg. 97477

Holbrook, Stewart. *Holy Old Mackinaw.* New York: Macmillan, 1956. A popular history of logging in the United States, written by a logger. Full of local color and anecdotal stories. Also a Ballantine paperback.

Jensen, Vernon H. *Lumber and Labor.* New York, 1945. An academic, de-romanticized picture of old-time loggers and their efforts to improve their conditions.

CHAPTER 2. TREE FARMING: THE VOICE OF INDUSTRY

Cleary, Brian D.; Greaves, Robert D.; and Hermann, Richard K. *Regenerating Oregon's Forests.* Corvallis, Oreg., 1978. An illustrated, practical explanation of the various tools and techniques used for commercial regeneration and intensive forest management. Write to:

Oregon State University Extension Service
Corvallis, Oreg. 97331

Hahn, Philip F. "How to Develop Super Trees." Eugene, Oreg., 1979. A succinct, practical outline for a commercial genetics program. Write to:

Manager, Forestry Research
Georgia-Pacific Corp.
P.O. Box 1618
Eugene, Oreg. 97401

The articles on commercial tree farming techniques are too numerous to cite individually. If you are interested in a specific field, consult the indexes of the forestry journals and the bibliographies put out by the Pacific Northwest Forest and Range Experiment Station of the Forest Service (see above). In addition to the forestry journals listed above, there are several magazines written specifically for the timber industry and generally available free-of-charge to workers and managers within the timber trade:

Forest Industries. Write to:
500 Howard St., San Francisco, Calif. 94105

Loggers Handbook. Write to:
The Pacific Logging Congress,
217 American Bank Building, Portland, Oreg. 97205

Logging Management. Write to:
300 W. Adams, Chicago, Ill. 60606

Timber/West. Write to:
P.O. Box 610, Edmonds, Wash. 98020

World Wood. Write to:
500 Howard St., San Francisco, Calif. 94105

CHAPTER 3. TREE SAVING: THE VOICE OF ECOLOGY

C.A.T.S. Newsletter. Current developments in the anti-herbicide movement. Write to:

Citizens Against Toxic Sprays
1385 Bailey Ave.
Eugene, Oreg. 97402

Coats, Robert N. "The Road to Erosion: A Cautionary Tale." *Environment* January-February 1978. A short, readable survey of the problems caused by deforestation and erosion.

CoEvolution Quarterly Winter 1976-1977. A special issue on watersheds, containing many brief selections on environmental forestry. Write to:

> POINT
> Box 428
> Sausalito, Calif. 94965

Dodge, Marvin; Burcham, L. T.; Goldhaber, Susan; McCulley, Bryan; and Springer, Charles. *An Investigation of Soil Characteristics and Erosion Rates on California Forest Lands.* Sacramento, Calif., 1976. If you want to study the effects of logging on watersheds, this is the place to start. There is a thorough review of the extensive literature on erosion and a comprehensive bibliography. Also included is a well-documented critique of present erosion hazard rating systems. Write to:

> California Department of Forestry
> 1416 Ninth St.
> Sacramento, Calif. 95814

Franklin, Jerry F.; Cromack, Kermit Jr.; Denison, William; McKee, Arthur; Maser, Chris; Sedell, James; and Swanson, Fred. *Ecological Characteristics of Old-Growth Forest Systems in the Douglas-fir Region.* Review draft of a Forest Service General Technical Report. A comprehensive ecological study that reveals the hidden benefits of snags and rotten logs.

Hermann, Richard K., and Lavender, Denis P., eds. *Even-Age Management.* Corvallis, Oreg., 1973. Probably the most far-reaching, yet thoroughly objective, academic scrutiny of clearcutting assembled in one volume. Included are a dozen evaluations of even-age management by experts from such diverse fields as wildlife management, hydrology, and microbiology. Write to:

> School of Forestry
> Oregon State University
> Corvallis, Oreg. 97331

Kittredge, Joseph. *Forest Influences.* New York: McGraw-Hill, 1948. A classic work on forests and the environment. Technical, but readable, information on all sorts of interrelated forest variables: filtering of light, air temperature, wind velocity, atmospheric moisture, precipitation, soil moisture, soil temperature, decomposition of litter, interception, transpiration, evaporation, surface runoff, stream flow, floods, and erosion. Available in a Dover paperback.

Lantz, Richard L. *Guidelines for Stream Protection in Logging Operations.* Portland, Oreg., 1971. A succinct and readable pamphlet explaining the relationship between logging and fisheries. Write to:

> Oregon State Game Commission

P.O. Box 3503
Portland, Oreg. 97208

Shoecraft, Billie. *Sue the Bastards*. Phoenix, 1971. A lively account of a personal confrontation with the Forest Service over the use of phenoxy herbicides.

Silen, Roy. "The Care and Handling of the Forest Gene Pool." *Pacific Search* June 1976. A professional geneticist talks about the possible dangers of bypassing natural selection.

Smith, Frederick E. "Ecological Demand and Environmental Response." *Journal of Forestry* December 1970. An eloquent statement by an eminent forester on behalf of the maintenance of natural balancing mechanisms.

Warnock, John W., and Lewis, Jay. *The Other Face of 2,4-D*. Penticton, B.C., 1978. A comprehensive review of the scientific literature on the health effects of 2,4-D, with some mention of the other phenoxy herbicides. The bibliography includes some 300 references to technical journals and governmental reports. Write to:
South Okanagan Environmental Coalition
P.O. Box 188
Penticton, B.C. V2A 6K3

CHAPTER 4. TREE GROWING: THE VOICE OF HOLISTIC FORESTRY

Forest Planning. A new journal put out by holistic forestry advocates. Write to:
C.H.E.C.
P.O. Box 3479
Eugene, Oreg. 97403

Groundwork, Inc. "Willamette Brush Control Study Project." Eugene, Oreg., 1978; and "Report on Hand Release Contracts, 1978." Eugene, Oreg., 1979. Groundwork is a research group composed of reforestation workers who are seeking safe, environmentally sound alternatives to the extensive use of chemicals on forested land. Many of the workers come from academic backgrounds, and their research is thorough, though directed toward preconceived goals. Write to:
Groundwork, Inc.
454 Willamette St.
Eugene, Oreg. 97401

The prospect of using green-manure trees instead of commercial fertilizers is receiving considerable attention in academic circles. For anyone interested in trying it out, here is the scientific background:

Bergstrom, Dorothy. "Let's Harness the Energy of Red Alder." *Forestry Research West* August 1979.

Briggs, D. G.; DeBell, D. S.; and Atkinson, W. A.; eds. *Utilization and Management of Alder.* Forest Service General Technical Report PNW-70. Portland, Oreg., 1978.

Gordon, J. C.; Wheeler, C. T.; and Perry, D. A.; eds. *Symbiotic Nitrogen Fixation in the Management of Temperate Forests.* Corvallis, Oreg., 1979.

"Red Alder, Once Wasted, Is Valuable Fuel Source." *Forest Industries* February 1980.

Tarrant, Robert F. "Stand Development and Soil Fertility in a Douglas-fir — Red Alder Plantation." *Forest Science* September 1961.

Tarrant, Robert F.; Lu, K. C.; Bollen, W. B.; and Franklin, J. F. *Nitrogen Enrichment of Two Forest Ecosystems by Red Alder.* Forest Service Research Paper PNW-76. Portland, Oreg., 1969.

Tarrant, Robert F.; Lu, K. C.; Bollen, W. B.; and Chen, C. S. *Nutrient Cycling by Throughfall and Stemflow Precipitation in Three Coastal Oregon Forest Types.* Forest Service Research Paper PNW-54. Portland, Oreg., 1968.

Trappe, J. M.; Franklin, J. F.; Tarrant, R. F.; and Hansen, G. M.; eds. *Biology of Alder.* Forest Service (PNW). Portland, Oreg., 1968.

Worthington, N. P.; Ruth, R. H.; and Matson, E. E. *Red Alder: Its Management and Utilization.* Forest Service Misc. Publication 881. Washington, D.C., 1962.

For information on energy uses and the possible implications for forestry, see: *American Forests* October 1978. This is a special issue on "Wood for Energy." Another good article on the subject is Norval Morey's "Clean the Land, Fuel the Country" in *Logging Management* June 1980.

CHAPTER 5. HARVEST TECHNOLOGY

Conway, Steve. *Logging Practices: Principles of Timber Harvesting Systems.* San Francisco: Miller Freeman Publications, 1976. A thorough explanation of the various logging techniques from an industrial perspective. It's an expensive hardcover, but worth the money if you are trying to learn practical strategies for logging. Write to:
    Forest Industries, Book Department
    500 Howard St.
    San Francisco, Calif. 94105

Cromack, K. Jr.; Swanson, F. J.; and Grier, C. C. "A Comparison of Harvesting Methods and their Impact on Soils and Environment in the Pacific Northwest." *Forest Soils and Land Use.* Edited by Chester T. Youngberg. Colorado State Univ., 1979. Insightful comparisons of the effects of various harvesting techniques on subsequent vegetation, nutrient depletion, soil disturbance, compaction, and erosion.

Journal of Forestry September 1973. An excellent series on logging methods: Hilton H. Lysons and Roger H. Twito's "Skyline Logging: An Economical Means of Reducing Environmental Impact of Logging"; Doyle Burke's "Helicopter Logging: Advantages and Disadvantages Must Be Weighed"; and Penn A. Peters's "Balloon Logging: A Look at Current Operating Systems." Clear, concise, and objective articles.

Klock, Glen O. "Impact of Five Postfire Salvage Logging Systems on Soils and Vegetation." *Journal of Soil and Water Conservation* March-April 1975. A comparison of soil disturbance in various logging operations, concluding that helicopters, skylines, or tractors operating on snow do far less damage to the soil than cable skidding or tractors operating on bare ground.

CHAPTER 6. SILVICULTURAL SYSTEMS

Bashline, Jim. "The Case for Clearcutting." *Field and Stream* July 1976. A popular account of the basic arguments for clearcutting: it's clean, it's efficient, and it provides maximum opportunities for engineered reforestation.

Daniel, Theodore W.; Helms, John A.; and Baker, Frederick C. *Principles of Silviculture.* New York, 1979. The basic college textbook on silviculture, explaining the nutritional and environmental requirements for tree growth.

McDonald, P. M.; Atkinson, W. A.; and Hall, D. O. "Logging Costs and Cutting Methods in Young Growth Ponderosa Pine in California." *Journal of Forestry* February 1969. A careful study of logging expenses involved in different silvicultural systems, showing but minor variations in overall costs.

Robinson, Gordon. "The Sierra Club Position on Clearcutting and Forest Management." Sierra Club Policy Paper No. 2. San Francisco, Calif. A clear and cogent presentation of the environmental virtues of selective cutting.

"The Uncommon Forest Management Plan." *Timber/West* February 1978. Encouraging results from a unique variation of the selective system.

U.S.D.A. Forest Service. *Silvicultural Systems for the Major Forest Types of the United States.* Agricultural Handbook 445. Washington, D.C., 1973. A succinct analysis of the viability of the common silvicultural systems for each of thirty-seven major forest types.

Williamson, Richard L. *Results of Shelterwood Harvesting of Douglas-fir in the Cascades of Western Oregon.* Forest Service Research Paper PNW-161. Portland, Oreg., 1973. A report of successful regeneration of Douglas-fir under the shelterwood method.

Williamson, Richard L., and Ruth, Robert H. *Results of Shelterwood Cutting in Western Hemlock.* Forest Service Research Paper PNW-201. Portland, Oreg., 1976. Regeneration was too successful: the sites became overly stocked.

CHAPTER 7. THE POLITICS OF TIMBER

Alston, Richard M. *Forest: Goals and Decisionmaking in the Forest Service.* Forest Service Research Paper INT-128. Ogden, Utah, 1972. A brief look at the Forest Service, including a history of Forest Service policy and an elaboration on the multiple-use philosophy.

Arvola, T. F. *Regulation of Logging in California: 1945-1975.* Sacramento, Calif., 1975. A case study of the progress, and the problems, in forest regulation.

Dana, Samuel T., and Johnson, Evert W. *Forestry Education in America: Today and Tomorrow.* Washington, D.C., 1963. A study of forestry schools in the United States, sponsored by the Society of American Foresters.

Robinson, Glen O. *The Forest Service.* Baltimore: Johns Hopkins, 1975. The definitive study of the Forest Service bureaucracy, written from the standpoint of a political scientist.

Wazeka, Bob. "Stashing Tomorrow's Trees: The Reforestation Boondoggle." *Willamette Valley Observer* 18 August 1978. An exposé of on-site corruption in the reforestation industry.

There have been several muckraking accounts of the timber industry and the Forest Service in the past decade:

Barney, Daniel R. *The Last Stand: Ralph Nader's Study Group Report on the National Forests.* New York: Grossman, 1974. This is the least biased and most informative of the exposés.

Shepherd, Jack. *The Forest Killers.* New York, 1975. A long and tedious account of timber politics.

Wood, Nancy. *Clearcut: The Deforestation of America.* San Francisco: The Sierra Club, 1971. A polemical work, but it contains interesting interviews with loggers and industrial foresters.

CHAPTER 8. LANDED FORESTRY: A VISION FOR THE FUTURE

Bernstein, Paul. "Run Your Own Business: Worker-Owned Plywood Firms." *Working Papers for a New Society* Summer 1974. The cooperative structure has worked well in the manufacturing end of the timber trade for forty years.

Eckholm, Erik. *Planting for the Future: Forestry for Human Needs.* Washington, D.C., 1979. Included in this apocalyptic report of worldwide deforestation are some inspirational stories of reforestation projects in underdeveloped countries. Write to:

> Worldwatch Institute
> 1776 Massachusetts Ave., N.W.
> Washington, D.C. 20036

Index

Also available from Island Press,
Star Route 1, Box 38, Covelo, California 95428

An Everyday History of Somewhere, by Ray Raphael. Illustrations by
Mark Livingston. $8.00
This work of history and documentation by the author of *Tree Talk*
embraces the life and work of ordinary people, from the Indians who
inhabited the coastal forests of northern California to the loggers, tan-
bark strippers, and farmers who came after them. This loving look at
history takes us in a full circle that leads to the everyday life of us all.

A Citizen's Guide to Timber Harvest Plans. Illustrations. $1.50
California state law permits any interested citizen to learn the details of
proposed timber cutting on private or public lands. This report by
Marylee Bytheriver instructs citizens on their rights concerning timber
harvesting, the procedures for influencing the details of proposed logging operations, and the specialized vocabulary surrounding the Timber Harvest Plan. A resource for action.

Wellspring: A Story from the Deep Country, by Barbara Dean. Illustrations. $6.00
The moving, first-person account of a contemporary woman's life at the
edge of wilderness. Since 1971, Barbara Dean has lived in a handmade
yurt on land she shares with 15 friends. Their struggles, both hilarious
and poignant, form the background for this inspiring story of personal
growth and deep love for nature.

*The Book of the Vision Quest: Personal Transformation in the Wilder-
ness,* by Steven Foster with Meredith E. Little. Photographs. $10.00
The inspiring record of modern people enacting an ancient, archetypal
rite of passage. This book shares the wisdom and the seeking of many
persons who have known the opportunity to face themselves, their fears
and their courage, and to live in harmony with nature through the experience of the traditional Vision Quest. Excerpts from participants'
journals add an intimate dimension to this unique account of human
challenge.

The Trail North, by Hawk Greenway. Illustrations. $7.50
The summer adventure of a young man who traveled the spine of coastal mountains from California to Washington with only his horse for a
companion. The book he has made from this journey reveals his coming
of age as he studies, reflects, and greets the world that is awakening
within and around him.

*Building an Ark: Tools for the Preservation of Natural Diversity
Through Land Protection,* by Phillip M. Hoose. Illustrations. $12.00
The author is national protection planner for The Nature Conservancy,

and this book presents a comprehensive plan that can be used within each state to identify and protect what remains of each area's natural ecological diversity. Case studies augment this blueprint for conservation.

Headwaters: Tales of the Wilderness, by Ash, Russell, Doog, and Del Rio. Preface by Edward Abbey. Photographs and illustrations. $6.00
Four bridge-playing buddies tackle the wilderness — they go in separately, meet on top of a rock, and come out talking. These four are as different as the suits in their deck of cards, as ragged as a three-day beard, and as eager as sparks.

Perfection Perception, with the Brothers O. and Joe de Vivre. $5.00
Notes from a metaphysical journey through the mountains of Patagonia. These two authors share their experiences and discoveries in using their powers of perception to change the world. Their thoughts are mystical at times, but their basis is firmly experiential and parallels the most theoretically advanced works in modern physics.

The Search for Goodbye-to-Rains, by Paul McHugh. $7.50
Steve Getane takes to the road in an American odyssey that is part fantasy and part real — a haphazard pursuit that includes Faulkner's Mississippi, the rarefied New Mexico air, and a motorcycle named Frank. "A rich, resonant novel of the interior world. Overtones of Whitman, Kerouac." — Robert Anton Wilson

No Substitute for Madness: A Teacher, His Kids & The Lessons of Real Life, by Ron Jones. Illustrations. $8.00
Seven magnificent glimpses of life as it is. Ron Jones is a teacher with the gift of translating human beauty into words and knowing where to find it in the first place. This collection of true experiences includes "The Acorn People," the moving story of a summer camp for handicapped kids that has been adapted for a television movie, and "The Third Wave," a harrowing experiment in Nazi training in a high school class — also a television special for 1981.

The Christmas Coat, by Ron Jones. Illustrations. $4.00
A contemporary fable of a mysterious Christmas gift and a father's search for the sender, which takes him to his wife, his son, and his memories of big band and ballroom days.

Please enclose $1.00 with each order for postage and handling.
California residents add 6% sales tax.
A catalog of current and forthcoming titles is available free of charge.